Y0-BCV-201

Renner Learning Resource Center
Elgin Community College
Elgin, IL 60123

POLYMERS FROM THE INSIDE OUT

POLYMERS FROM THE INSIDE OUT

An Introduction to Macromolecules

ALAN E. TONELLI
with
MOHAN SRINIVASARAO

WILEY-INTERSCIENCE

A John Wiley & Sons, Inc., Publication

New York • Chichester • Weinheim • Brisbane • Singapore • Toronto

RENNER LEARNING RESOURCE CENTER
ELGIN COMMUNITY COLLEGE
ELGIN. ILLINOIS 60123

54'7.7
T66⁴ᵖ

This book is printed on acid-free paper.⊗

Copyright © 2001 by John Wiley & Sons, Inc. All rights reserved.

Published simultaneously in Canada.

No part of this publication may be reproduced, stored in a retrieval system or transmitted in any form or by any means, electronic, mechanical, photocopying, recording, scanning or otherwise, except as permitted under Section 107 or 108 of the 1976 United States Copyright Act, without either the prior written permission of the Publisher, or authorization through payment of the appropriate per-copy fee to the Copyright Clearance Center, 222 Rosewood Drive, Danvers, MA 01923, (978) 750-8400, fax (978) 750-4744. Requests to the Publisher for permission should be addressed to the Permissions Department, John Wiley & Sons, Inc., 605 Third Avenue, New York, NY 10158-0012, (212) 850-6011, fax (212) 850-6008, E-Mail: PERMREQ@WILEY.COM.

For ordering and customer service, call 1-800-CALL-WILEY.

Library of Congress Cataloging-in-Publication Data:

Tonelli, Alan. E.
    Polymers from the inside out : an introduction to macromolecules / Alan E. Tonelli with Mohan Srinivasarao.
        p.   cm.
    Includes index.
    ISBN 0-471-38138-1 (cloth : alk. paper)
    1. Polymers. 2. Polymerization. I. Srinivasarao, Mohan. II. Title.
QD381.T65   2001
547'.7—dc21   00-047990

Printed in the United States of America.

10 9 8 7 6 5 4 3 2 1

CONTENTS

PREFACE

Polymers or macromolecules dominate our world. From naturally occurring polymers necessary to sustain life, such as proteins, poly-nucleotides (DNA, RNA), and polysaccharides (cellulose, starch, chitin, etc.), to the polymers of commerce (polyethylene, polystyrene, polyvinylchloride, nylons, polyesters, rubbers, etc.), macromolecules are the building blocks of choice used to construct the materials found in both the natural and man-made worlds. The dominance of polymeric materials is a consequence of their unique physical properties that result from their long-chain structures and high molecular weights.

These two unique structural features permit individual polymer chains to assume an almost limitless array of sizes and shapes in response to both the environments in which they are placed and the forces to which they are subjected. It is the large sizes and variable shapes which macromolecules can so readily adopt that distinguish them from all other forms of matter and lead to their unique physical properties.

The one-dimensional, long-chain nature of polymers not only results in their ability to respond internally to different environments and stresses, something that small-molecules, atoms, and ions cannot do, but also facilitates an understanding of their properties from a molecular-level point of view.

Backbone bonds in most polymers are easily rotated generally between their preferred staggered conformations, thereby generating a myriad of overall polymer conformations $N_{conf} = C^n$, where $n =$ the number of backbone bonds and $C =$ the number of preferred conformations adopted by each backbone bond. Furthermore, the energy associated with any particular backbone bond conformation usually depends only on the conformations of the backbone bond in question and those of its immediate neighbors. For this reason the overall or total energy of a particular polymer chain conformation E_{conf} may be simply estimated through the summation of pairwise-dependent bond conformational energies $E(\phi_{i-1}, \phi_i)$. $E_{conf} = \sum_{i=2}^{n-1} E(\phi_{i-1}, \phi_i)$, where $E(\phi_{i-1}, \phi_i)$ corresponds to the local conformational energy when bonds $i - 1$ and i adopt the conformations ϕ_{i-1} and ϕ_i.

If the collection of polymer chain conformations N_{conf} is considered an ensemble of statistical mechanical systems, with each system of the ensemble identified with a particular overall polymer chain conformation, then we may readily evaluate the conformational partition function of the polymer chain and all its attendant thermodynamic properties, because the system energies, E_{conf}, are readily estimable for polymer chains.

Application of matrix multiplication techniques developed for one-dimensional statistical mechanical systems permits the rigorous evaluation of both global and local properties of a single polymer chain, such as its end-to-end distance (size) and local backbone bond conformational populations, respectively, which are appropriately averaged over all of its N_{conf} conformations. This approach accounts for in a realistic way the detailed microstructural features of polymers which distinguish, for example, polyethylene from polystyrene.

The conformational characteristics of individual, isolated polymer chains and their resulting overall sizes and shapes may in many cases be related to their unique physical properties as manifested in both dilute solution and pure bulk states, where the polymer chains are no longer isolated from each other or from other molecules. When successful, this approach to understanding the behavior of macromolecules yields a molecular-level knowledge of their structure–

property relations, which remains the "Holy Grail" in materials science.

Several physical properties unique to polymeric materials are introduced in Chapter 1 and are qualitatively related to their high-molecular-weight, long-chain structures to distinguish them from small-molecule and atomic systems. Step-growth and chain-growth polymerizations are described in Chapters 2 and 3 without introducing more than the necessary elements of organic chemistry. The microstructures of polymers, principally chain-growth vinyl polymers, are introduced and discussed in Chapter 4, including the microstructures of copolymers, branching, and cross-linking. Chapter 5 covers the conformational characteristics of polymers developed with the rotational isomeric states (RIS) model and introduces in an elementary way the matrix multiplication methods necessary for the calculation of polymer chain properties that are appropriately averaged over all of their conformations.

Chapters 6 and 7 describe the solution and bulk properties of polymers with emphasis placed on their connections with the conformational characteristics of individual polymer chains. We close with a chapter (Chapter 8) on biopolymers, where the emphasis is placed on the connections between their microstructures (primary structures), conformations (secondary structures), overall sizes and shapes (tertiary structures), and properties (biological functions). Here is presented the inescapable conclusion that life is not possible without polymers.

In the early portion of each chapter, topics are intoduced and elementary examples and illustrations of the requisite concepts and methods are provided. In the remaining portion of each chapter these topics, concepts, and methods are elaborated upon, and sometimes additional new material is introduced and discussed. Organization of material in this format is designed to facilitate its use in teaching either a one- or two-semester undergraduate course in polymer science. In the one-semester course only the early portions of each chapter need be covered, while in the second semester of the two-semester sequence course the initial material in each chapter may be quickly reviewed with the remaining material serving as the focus of coverage in each chapter.

Simple classroom demonstrations appropriate to the material

covered in each chapter are described, and several questions raised by each demonstration are posed to generate student inquiry. Also a set of homework questions and problems are provided at the end of each chapter, and those believed appropriate for second-semester students are denoted with an asterisk.

We take the above approach to presenting the covered material, because we believe that all students should be exposed to the topics covered in each and every chapter, with expanded, more in-depth coverage of the same topics for those students fortunate to have access to a two-semester course in polymer science.

ALAN E. TONELLI
MOHAN SRINIVASARAO

ACKNOWLEDGMENTS

To the late Paul J. Flory, who introduced me to the fascinating world of macromolecules and who, by many careful examples, instilled in me a chemist's molecular point of view concerning their behavior, I am grateful. In addition, for my continuing enthusiasm for polymer science I must thank my collaborators and colleagues of the past 30 years.

ALAN E. TONELLI

CHAPTER SUMMARY

In Chapter 1 we define polymers/macromolecules (long-chain molecules of high molecular weight) and qualitatively describe several of their unique properties, such as dilute solution viscosities, rubber elasticity, and time-dependent responses. (Each of these can be introduced using the appropriate demo.) These behaviors unique to polymers are contrasted to the responses of small-molecule and atomic systems subjected to the same environments and stimuli, and the differences are attributed to the physical distinctions between macromolecules and all other materials—that is, the high molecular weights, large sizes, and variable shapes that are only possible for long-chain molecules. Small-molecule and atomic systems can only respond to stimuli (environment or stresses) by changing the arrangements of their constituents as manifested by the distances between and orientations among their molecules or atoms. Polymers, on the other hand, have an additional internal degree of freedom, their variable sizes and shapes, provided by their ability to alter their conformations *via* internal rotations about their backbone bonds. It is this additional internal degree of freedom which permits polymer chains to respond uniquely from the "inside" to various stimuli, resulting in a unique set of "outside" physical properties. It is the physics of polymers, their high molecular weights, long-chain structures, and variable sizes and shapes which distinguish them from all other materials. There is nothing about their constituent atoms and groups which is unique from other materials, so the chemistry of polymers, as reflected in their different microstructures, simply serves to distinguish behaviors within their own macromolecular class of materials.

Three demonstrations that emphasize the unique behaviors or responses of polymers and their differences from the behaviors or responses of materials composed of small-molecules or atoms should be performed in this introductory chapter. We begin by comparing the flow times of water and 1 wt% aqueous solutions of ethylene glycol (EG) and poly(ethylene oxide) (PEO) as observed in a Cannon–Ubbelhode viscometer, employing a PEO sample with a high molecular weight (4,000,000). The flow time for the PEO solution, and therefore its viscosity, is several times the flow times observed for the EG solution and for pure water. At this point we can ask the students to speculate or hypothesize how such a small amount of the high-molecular-weight PEO can so dramatically slow the flow of water.

Observations made by each student as they stretch a rubber band illustrates several unique aspects of elastic polymer networks: (1) reversible, high extensibility, (2) low modulus, (3) extension of the rubber band in the stretching direction and contraction in the thickness and width directions, and (4) warming/cooling of the rubber band during stretching/contraction, as detected by placing the rubber band between the inner lips. Again we suggest that students be encouraged to offer reasons for these unusual behaviors.

Observation of the time-dependent responses of slime [5% poly(vinyl alcohol) in water to which is added a few drops of a 5% borax (Na_2BO_3 solution] offers a means to impress upon students the fundamentally different responses of polymeric systems. Form a ball and a cylinder of the slime, bounce the slime ball on a hard counter top, bend the cylinder over the edge of the counter top, and pull on the cylinder with varying amounts of force applied for different durations. The slime will respond as either a viscous liquid, an elastic solid, or a brittle solid depending on the magnitude and duration of the perturbing force. By comparison to the behavior of liquid and frozen water, the slime demonstration can be used to generate considerable discussion concerning the source of the time-dependent properties of slime in particular and of polymeric systems in general.

In Chapter 2 we describe the linking of monomers containing at least two functional groups that react and eventually, via a slow, steady step-growth process, lead to polymer chains. The simultaneous reactivity of all functional groups is stressed; and the resultant production of small and medium-sized oligomers throughout most of the reaction, with reasonably high molecular weight polymer achieved only near the very end of the reaction if the monomer-to-polymer conversion is nearly complete, is described. The need to achieve high conversions to obtain reasonably long polymer chains is demonstrated with the Carothers equation and used to explain why only a handful of all the possible reactions between organic functional groups can be used to step-grow polymers. The functional groups appropriate to step-growing polymers are introduced along with their linking reactions and the resultant polymers.

In the latter part of this chapter the thermodynamics and kinetics of step-growth polymerizations are introduced. Here their consequences, such as the formation of cyclic oligomers, the necessary stoichiometric equivalence of functional groups, the assumptions embodied in and the experimental justification for the equal reactivity of functional groups, and the need to conduct most step-growth polymerizations in open systems, so the byproducts (H_2O or HCl) of monomer condensation can be removed to allow their reaction to be driven to the requisite level of completion, are presented.

A demonstration is utilized to convey the stepwise buildup and consequent time dependence of step-growth polymerizations. It is called "Polymerization in a Box" and requires a small box and 100 pieces of paper cut in 1-inch squares from index cards. We assume each piece of paper to represent a monomer of the type X–R–Y, where X,Y are functional groups that react with each other to form a –XY– link, but do not self-react. Their polymerization is begun by reaching into the box and removing two monomers, followed by discarding one and writing a 2 on the other to denote a dimer and returning it back to the box. This process is repeated and is always accompanied by discarding the piece of paper with the smaller number written on it, and then we reintroduce the piece of paper with the larger number back into the box after its number is increased to the sum of the numbers on the two pieces of paper as they were selected from the box. To ensure a random step-growth simulation, the box should be closed and shaken between reaction steps. During the course of this simulated step-growth polymerization, a record of the box contents should be kept, including how many monomers and oligomers of various lengths it contains. In this way we can follow the slow buildup of oligomers to eventually obtain a polymer of 100 repeat units and characterize the reaction products in terms of their number and weight averages and the longest chain produced as the conversion of the X,Y functional groups proceeds.

The initiated chain-growth polymerization of unsaturated monomers of the predominantly vinyl type is presented in Chapter 3. We begin by discussing the potential reactivity of the two π electrons in the sp^2–sp^2 C=C bond(s) of the monomers and point out the need for activating the monomer by reaction with an initiator. Free-radical initiation is stressed here, though ionic initiators are mentioned later in the chapter when living polymerizations are described. Once a monomer is activated by the initiator, rapid monomer addition follows to produce the chain-growth polymer. Occasional termination of growing chains via coupling, disproportionation, and chain-transfer are described and their consequences regarding the structures and molecular weights of the chain-growth polymers are developed. The differences between chain-growth and step-growth polymerizations are pointed out with emphasis placed on the times required to obtain high-molecular-weight polymers and their structures.

In the latter portion of this chapter we examine the kinetics of chain-growth polymerization to learn how the molecular weight and overall polymerization rate are affected by the monomer and initiator concentrations and by how the growing chains are terminated. The elements of emulsion and ionically initiated living polymerizations are also introduced and briefly developed here.

Comonomer compositions and sequence distributions introduced during the copolymerization of two monomers are also treated, including the concept of relative reactivity ratios and the resultant copolymer equation of Mayo. Emphasis is placed on using the copolymer equation and the relative reactivity ratios for comonomer pairs to predict the copolymer composition and comonomer sequence distribution.

The chain-growth polymerization of polyacrylamide initiated with riboflavin offers a valuable demonstration of several aspects of this class of polymerizations. A 50 wt% aqueous solution of acrylamide monomer is placed in a small beaker, and a few drops of an aqueous solution of riboflavin (vitamin B_2) are added. Nothing happens to the mixture of solutions, until the beaker is placed on the glass surface of an overhead projector where it is exposed to UV-VIS light of the appropriate wavelengths to cause the riboflavin to photo-initiate free radicals (OH$^{\cdot}$). As these free radicals initiate several chains, which quickly grow through the rapid addition of acrylamide monomer, the reaction solution begins to sputter and become noticeably warmer with the evolution of steam, because of the highly exothermic nature of the addition of acrylamide monomer to the growing polyacrylamide chains. As the amount of polyacrylamide produced increases, the viscosity of the solution increases, eventually leading to a leathery disc of water-swollen polyacrylamide containing trapped bubbles of the evolving steam which are clearly visible. The speed and exothermic nature of this chain-growth polymerization are readily apparent, as is the need to expose the riboflavin to light of the appropriate wavelengths in order to produce the initiating free radicals.

The interfacial polymerization of sebacyl chloride and hexamethylene diamine to yield nylon-6,10 along with the polyacrylamide chain-growth polymerization and "polymerization in a box" demos, provides a nice comparison of both classes of polymerizations. In this familiar demonstration each bifunctional monomer is dissolved in different immiscible solvents, so that a very rapid polymerization is confined to the interface between the two immiscible monomer solutions. This quick production of a traditional step-growth polymer contrasts nicely with the slow progressive polymerization conducted in the box, and it reinforces the concepts of equal reactivity of functional groups, stoichiometric equivalence of reacting functional groups, the Carothers equation, and the reactivities of chain- and step-growth monomers.

Microstructural elements such as regiosequence, stereosequence, geometrical isomers, and mode of addition, which are irreversibly incorporated into a chain-growth polymer during the propagation, termination, and chain-transfer phases of growing the polymer, are presented in Chapter 4. These elements are combined with the composition and comonomer sequence distribution to cover the microstructures of copolymers. Finally the introduction of branches and cross-links along the polymer backbone are mentioned.

A very brief introduction to the ^{13}C NMR observation and determination of polymer microstructures is presented, where the necessary connections between the local polymer microstructures, the local polymer conformations, and the resonant frequencies of the carbons in the local polymer microstructures are pointed out. In addition, a very preliminary comparison of polymer microstructures and polymer properties is made via several specific examples, such as isotactic versus atactic polypropylene and linear versus branched polyethylenes.

The purpose of Chapter 5 is to describe the conformational characteristics of polymers, which govern their overall sizes and shapes and thus their responses to environmental stimuli. We stress the connections between a polymer's ability to undergo facile backbone rotations, which are driven by the desire to maximize its entropy and to increase its favorable interactions with the environment, and its overall average size and variable shapes, because they may be related to the macroscopic behavior of polymeric materials. The chemical structure of the polymer is explicitly considered when estimating the likelihood of its various local conformations, and so our development is designed to permit a distinction between the conformational properties of chemically different polymers. The rotational isomeric state (RIS) model of polymer conformations is detailed, and emphasis is placed on the immense number of conformations and the vast range of overall sizes and shapes that are available to polymers, even if we restrict their backbone bonds to adopt only a small number, usually 3, of staggered rotational, conformational states.

We indicate how the local conformational characteristics of a polymer as embodied in its RIS model may be used to calculate properties that depend on averaging over all of its myriad conformations. The distance between the ends of a polymer chain, $\mathbf{r}$, is utilized as a means to characterize its overall or global size, and means for calculating $\langle \mathbf{r}^2 \rangle$ by properly averaging over all available conformations are outlined. We utilize $\langle \mathbf{r}^2 \rangle$ to characterize the average volume or environmental expanse that a polymer will influence as it samples its many conformations. This volume, V_i, is contrasted with V_0, which is the physical or hard-core volume occupied by its constituent atoms and groups. It is noted that V_i/V_0 is ~ 100, so the volume influenced by a polymer free to adopt its many possible conformations far exceeds the volume occupied by its segments and excluded to all other molecules. Following this development, we then explain why small amounts of polymers greatly increase the viscosities of their solutions and why in disordered, bulk polymer samples each polymer chain interacts/entangles with many other polymer chains.

We will also point out the large range of sizes ($\langle \mathbf{r}^2 \rangle$) available to high-molecular-weight polymers. Comparison of $\langle \mathbf{r}^2 \rangle$ for a dissolved or molten randomly coiling polymer chain to its fully extended length makes clear that a polymer's size can vary over nearly two orders of magnitude, which leads to unique behaviors such as rubber elasticity and viscoelasticity.

An exercise that reinforces the statistical nature of polymer conformations involves calculation of the partition function, Z_{conf}, of an isolated *n*-hexane molecule. This is first accomplished by hand from the E_σ and E_ω energies, which correspond to the first- and second-order interactions occurring in *n*-butane and *n*-pentane. Then the statistical weight matrices describing the three rotations about the internal bonds in *n*-hexane are developed and used in the matrix multiplication scheme to evaluate Z_{conf} directly. Bond conformational populations are also calculated both by "hand" and with the matrix multiplication techniques. This exercise, which is more appropriate for second-semester students, should give the impression that there are rigorous methods for treating the conformational characteristics of polymers, and these may be extended to treat global properties, such as the physical size of the polymer chain, which must be averaged over all possible conformations.

We begin Chapter 6 with a discussion of how much volume a randomly coiling polymer chain influences as it assumes its myriad conformations in solution with reference to the introduction of this topic in Chapter 5. Here we point out that our esimate of V_i assumes a size ($\langle \mathbf{r}^2 \rangle$) given by the RIS treatment of polymer conformations, which is limited to the consideration of short-range, intramolecular interactions (pairwise backbone rotations). However, in solution the segments of a polymer are not only interacting with each other, but they are in intimate contact with solvent molecules. We introduce the concept of excluded volume expansion of polymer chain conformations above the unperturbed value $\langle \mathbf{r}^2 \rangle_0$ obtained from the RIS model, and Flory's idea of a θ-temperature for a particular polymer-poor solvent system, which causes a contraction of the polymer chain to relieve polymer segment–solvent interactions and which exactly balances the expansion of the polymer due to excluded volume self-intersections.

At this point the expression for the intrinsic viscosity of a dilute polymer solution is derived from Einstein's equation describing the viscosity of a liquid containing small, impenetrable spheres, which is appropriate to dilute polymer solutions if each polymer random coil is non-free draining and therefore moves through the solvent as an impenetrable sphere with volume $= V_i$. The expression simplifies to $[\eta] \propto (\langle \mathbf{r}^2 \rangle_0^{3/2})/\overline{M}_v$ at $T = \theta$, so we can estimate the dimensions $\langle \mathbf{r}^2 \rangle_0$ or the molecular weight of a polymer from the measurement of its intrinsic viscosity in a θ-solvent, provided that we know either $\overline{M}_v$ or $\langle \mathbf{r}^2 \rangle_0$ independently.

Brief mention is made of using light scattering from dilute polymer solutions to obtain both $\overline{M}_w$ and $\langle \mathbf{r}^2 \rangle$ simultaneously, and GPC/SEC determination of polymer molecular weights will also be introduced. In this connection, number and weight averaging of molecular weights $(\overline{M}_n, \overline{M}_w)$ are described, and their relationship to the viscosity-average molecular weight, $\overline{M}_v$, is mentioned. To describe the breadth of the molecular weight distribution the polydispersity index $\text{PDI} = \overline{M}_w/\overline{M}_n$ is defined and discussed briefly for both step- and chain-growth polymers.

This chapter closes with mention and a brief discussion of the flow properties of polymer solutions, in both their Newtonian and viscoelastic, non-Newtonian shear flow regimes.

A demonstration of the effect of shear rate on the flow properties of polymer solutions can be conveniently carried out by observing the solvent and dilute polymer solution flow times in Ubbelhode viscometers with varying capillary diameters. The shear rates achieved in large-diameter viscometers will lead to a decrease in the ratio of solution to solvent flow times compared with the flow times observed in narrow capillary viscometers.

In Chapter 7 we begin our discussion with polymer fluids and their viscoelastic properties. This includes the shear and molecular weight dependences of melt viscosities and their relation to V_i/V_o. Here the importance of the viscoelastic behavior of polymers in relation to their processing and ultimate uses is stressed. Amorphous solid polymers are treated next, beginning with their glass-transition phenomena. Molecular parameters such as molecular weight, side-chain bulk and interactions, and inherent backbone chain flexibility are mentioned as factors potentially important in determining at what temperature (T_g) an amorphous solid polymer makes the transition from a glassy, brittle to a tough, leathery material, with an emphasis placed on the ability of a polymer to change (or not) its conformation above (below) T_g. Though we are currently unable to predict the T_g of a polymer based solely on its microstructure, we can rationalize the comonomer sequence-dependent T_g's observed for copolymers and the high impact strength observed for certain polymers well below their T_g values. Both of these features are understandable from a knowledge of the conformational characteristics (RIS) of polymers, and several examples of understanding these aspects of T_g behavior will be presented.

The behavior of elastic polymer networks achieved through the chemical cross-linking of mobile amorphous polymer fluids is next described, including the requirements necessary for low modulus, reversible, high extensibility. We present the simplified thermodynamic treatment of polymer elastic networks and demonstrate the entropic (conformational) origin of the restoring force. From a knowledge of the conformational characteristics of uncross-linked polymers, we demonstrate their connection to fundamental properties of its elastic cross-linked network, such as its modulus.

Crystalline polymers are discussed in terms of their conformations, morphologies, melting, and properties. It is pointed out that almost invariably, polymers crystallize in the lowest energy conformation available to their isolated chains, as indicated by their RIS analyses, without significant distortions arising from the close-packing of neighboring chains in their crystals. The lamellar morphologies of crystalline polymers arising from their kinetically controlled crystallization are reviewed, and the two or more phases (crystalline, amorphous, interfacial) that coexist in a bulk semicrystalline polymer sample are stressed. This morphology is related to their physical properties, with particular reference to the fabrication and strengthening of polymer fibers.

The melting of crystalline polymers is described with a view to understanding the factors important to their melting temperatures, which are so important to their commercial applications. Melting of polymer crystals is analyzed on the basis of a two-step process: (i) expansion of the polymer crystal to the density of the molten phase without conformationally disordering the extended polymer chains and (ii) constant-volume conformational disordering of the extended polymer chains to their randomly coiling melt conformations. The energy and entropy accompanying the second step in melting is estimated from the RIS conformational characteristics of polymers.

Comparison to the total enthalpy and entropy of melting reveals that ΔH_m and ΔS_m are dominated by the separation of chains and the conformational disordering of chains, respectively, which accompanies melting, so roughly $T_m = \Delta H_m(\text{i})/\Delta S_m(\text{ii})$. This analysis enables us to suggest that high-melting polymers are those with strong interchain interactions in their crystals and whose molten random coils are relatively inflexible. On the other hand, low-melting polymers have weak interchain interactions in their crystals and upon melting adopt flexible random-coiling conformations. This approach, which relies on a knowledge of the conformational characteristics of single chains (inside behavior), can be effectively employed to discuss an important physical characteristic of a bulk crystalline sample (outside behavior) such as its T_m.

Several demonstrations can be presented to illustrate the unique behaviors of bulk polymer samples. A rubber band placed between the lips can be detected to noticeably warm as it is extended; and if first extended and then allowed to contract between the lips, a cooling sensation results. Placement of a rubber band inside a glass tube and attaching a weighted pan to the enclosed rubber band allows us to observe its behavior when heated with a hair dryer while under tension (contraction and lifting of weighted pan). Both demos illustrate the unusual thermal behavior of elastic polymer networks under tension and, when coupled with elementary thermodynamic considerations, can identify the entropic/conformational origin of rubber elasticity. Placement of a small bar of plexiglass (PMMA) or an old pair of plastic eyeglass frames in a beaker of boiling water will soften and allow them to be twisted into different shapes that are retained after quenching both specimens into cold water. This demo provides a nice demonstration of T_g and may be coupled with the submerging of a stretched rubber band into liquid nitrogen and observation of its retraction, only after removal and warming, to point out the necessity of $T > T_g$ for polymer networks to exhibit reversible elasticity.

In Chapter 8 the important polysaccharides starch, cellulose, chitin, and chitosan are introduced, and their distinguishing physical properties are discussed in terms of their unique yet similar microstructures. Sizing and desizing of cotton (cellulose) with starch (amylose) nicely illustrates the remarkably different properties of these structurally similar polysaccharides.

In our discussion of proteins their primary, secondary, tertiary, and quaternary structures are established and related to the structural elements of synthetic polymers, such as their microstructures, local conformations, overall sizes and shapes, and morphologies. The regular conformations adopted by the structural or fibrous proteins are contrasted to the compact, (seemingly) irregular conformations of the globular proteins that function as enzymes, carriers, and regulators. We draw the parallel between the structures and the behaviors of nylons and the fibrous keratin proteins silk and wool, and their contrasting properties are identified with their structural differences. By means of contrast between the potentially large variety of possible proteins and the limited set of proteins required by any living organism to function effectively, the need for a polymer synthesis method that is both precise and efficient becomes apparent.

The polynucleotides DNA and RNA are identified as the means to achieve the necessary specificity in protein synthesis, and their structures are presented and related to this critical function. DNA replication, made possible by base pairing and the resultant double-helical structure, is described, and the requirements and identification of the gentic code are presented. Finally a highly pictorial description of DNA directed protein synthesis is presented, and the rudiments of genetic diseases and bioengineered proteins are illustrated.

Natural samples of starch, cellulose, chitin, wool, silk, and so on, offer the opportunity to see first hand the remarkable range of properties exhibited by biopolymers.

CHAPTER 1

INTRODUCTION

LONG-CHAIN MOLECULES

Our purpose here is to introduce you to polymers or macromolecules. We begin with the generic structure of polymers, which is simply a long chain-like arrangement of chemically bonded monomers. As an example, let us combine two different molecules, each containing a pair of functional groups reactive toward the other pair. Ethylene glycol (EG) (1,2-ethane diol), $HO-CH_2-CH_2-OH$, and adipic acid (AA) (1,6-hexanedioic acid), $HO-\overset{\overset{O}{\|}}{C}-CH_2-CH_2-CH_2-CH_2-\overset{\overset{O}{\|}}{C}-OH$, will react to form the following ester: $HO-\overset{\overset{O}{\|}}{C}-CH_2-CH_2-CH_2-CH_2-\overset{\overset{O}{\|}}{C}-O-CH_2-CH_2-OH$. Because both ends of this ester may react further with EG and AA, it is readily apparent that a long-chain polyester with the following structure will eventually be produced: $-(O-\overset{\overset{O}{\|}}{C}-CH_2-CH_2-CH_2-CH_2-\overset{\overset{O}{\|}}{C}-O-CH_2-CH_2-)_n-$, where n is the degree of polymerization and poly(ethylene adipate)(PEA) is the common name for this polyester. The molecular weight of PEA is

1

simply 172n, because the molecular weight of a PEA repeat unit is 172. Typically, n is of the order 100–200 for polyesters (see Chapter 2), so molecular weights of $\sim$20,000–40,000 are common. Now con-

sider the simple ester $CH_3-CH_2-CH_2-CH_2-\overset{\overset{\text{O}}{\|}}{C}-O-CH_2-CH_3$ (ethyl pentanoate, EP) made from pentanoic acid and ethanol. Its molecu-

lar weight is 130 and its constituent groups ($CH_3, CH_2, \overset{\overset{\text{O}}{\|}}{C}, O$) are connected by seven bonds. On the other hand, a typical PEA polymer will possess 10n or $\sim$1000–2000 bonds, each identical to those found in the simple ester. What distinguishes the liquid ester EP from the plastic film and fiber forming PEA are their disparate sizes. Though polyesters like PEA are among the smallest polymers known, they remain hundreds of times larger than their more familiar small-molecule counterparts like the EP ester.

An even more dramatic demonstration of the disparity between the sizes of small molecules and polymers or macromolecules is offered by the two-carbon alkane gas ethane (CH_3-CH_3) and its related polymer polyethylene (PE) $-(CH_2-CH_2-)_n-$, where n typically ranges from 10,000 to 100,000. If the ethane molecule was magnified a billion times, it would be $\sim$1 foot in length. A typical PE chain ($n = 25,000$), by comparison, could span a distance greater than 4 miles if it were completely extended. Thus, polymers are truly macromolecules with sizes that are orders of magnitude larger than the small molecules most chemists are familiar with. [The reader is encouraged to refer to Morawetz (1985) for an interesting and insightful historical development of the macromolecular hypothesis put forward by Staudinger and demonstrated by Carothers.]

The following demonstration provides a tangible example of the effects produced by the large sizes of polymers on their physical properties. We begin by making two aqueous solutions, one containing 1 wt% ethanol and the other 1 wt% of a high-molecular-weight poly(ethylene oxide) (PEO), $-(-CH_2-CH_2-O-)_n-$, with $n = 91,000$. Each of these solutions are placed in turn into a capillary viscometer as shown in Figure 1.1, and the time required for each solution to drain from the upper solution reservoir through the narrow capillary into the lower reservoir is recorded. These solution flow times are then compared to the time required for pure water to flow through the

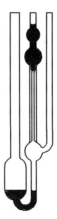

Figure 1.1. Cannon–Ubbelohde viscometer.

capillary. For a viscometer (#75) that requires a 2-minute flow time for water, the flow times for the 1 wt% ethanol and PEO ($n = 91,000$) solutions are roughly 2 and 90 minutes, respectively. Whereas the flow time for the 1 wt% ethanol solution is nearly the same as pure water, the flow time observed for the 1 wt% PEO solution is more than 40 times as long as that for pure water. Though each solution contained 1 wt% of solute, the numbers of ethanol and PEO molecules in each solution are very different. For PEO with $n = 91,000$ and a molecular weight of 4,000,000 and ethanol with a molecular weight of 46, the ratio of ethanol/PEO molecules in each of their 1 wt% solutions is $4,000,000/46 = 87,000$. Clearly the much larger size of the fewer PEO molecules overwhelmingly retards the flow of water compared with the much smaller though much more numerous ethanol molecules. We will in Chapter 6 return to the general observation that very small amounts of high-molecular-weight polymers can lead to very large increases in their solution viscosities. (The viscosity of a liquid is inversely related to its rate of flow.) There we will develop the connection between the physical size of a dissolved polymer and the retarding effect it has on the flow of its dilute solutions.

CONFORMATIONS, SIZES, AND SHAPES

Recall that a typical PE chain, $n = 25,000$ and molecular weight $= 700,000$, is, when fully extended, over 20,000 times as long as the chemically related small molecule ethane, CH_3-CH_3. Just as the methyl groups in ethane can be easily rotated (see Figure 1.2), so too is it possible for PE to change its conformation through rotations about its backbone bonds. By way of the examples presented in Figure 1.2, we can clearly see that backbone bond rotations in polymer chains can dramatically effect their sizes and shapes by altering their conformations. To the contrary, ethane and most other small molecules are not able to significantly change their sizes and shapes through adjustment of their conformations. In conjunction with their large sizes, the ability of polymers to change their conformations, thereby dramatically altering both their sizes and shapes, provides polymer chains with an internal degree of freedom not available to small-molecule or atomic materials.

Before we can expect to understand the reasons why polymeric materials often exhibit such unique physical properties, we must appreciate that their long-chain natures confer upon them the ability to assume a vast array of overall sizes and shapes that are achievable by backbone bond rotations and which lead to conformations as different as those shown in Figure 1.2(c) and (d). It is not too difficult to imagine that a solution or a bulk sample of PE whose chains are restricted exclusively to either the extended or the folded conformations depicted in Figure 1.2(c) and (d) might behave quite differently. Certainly the interactions between an extended or a folded chain with solvent molecules in solution or with other like chains in the bulk would be expected to differ and produce macroscopically different physical behaviors. As a consequence, we must concern ourselves with the conformational characteristics of the individual, constituent polymer chains in a macroscopic sample before we can begin to understand the physical properties they manifest. Hence, this is why we adopt the *inside/out* point of view and approach to polymer science.

The defining ability of polymers to significantly alter their sizes and shapes by adjusting their conformations through backbone bond rotations in response to an environmental stimulus can be conveniently illustrated with a rubber band. In Chapter 7 we will examine more closely the structures and properties of elastic polymer networks, such as rubber bands. But for the moment, it is sufficient to say that the *cis*-
$$CH_3$$
$$|$$
1,4-polyisoprene (c-PIP), $-(CH=C-CH_2-CH_2-)_n-$, chains in a rubber band are permanently attached to one another through occasional

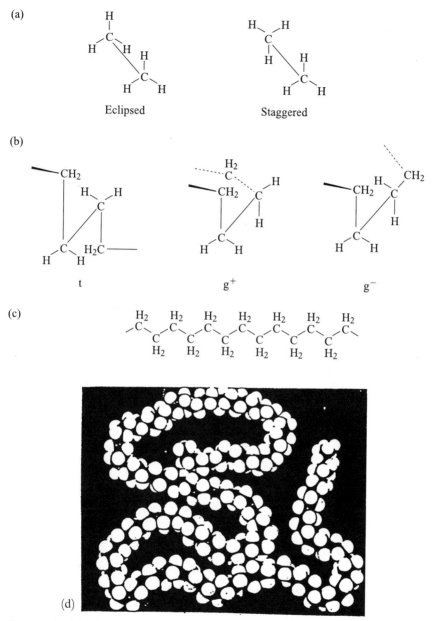

Figure 1.2. (a) Newman diagrams of ethane conformers. (b) Newman diagrams of trans (t) and gauche (g±) conformers in PE. (c) PE in the extended all trans and (d) a folded conformation.

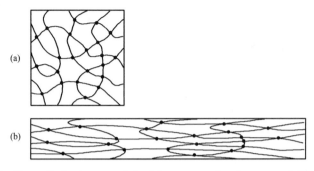

Figure 1.3. Cross-linked network in (a) an unstrained state and (b) a strained state.

C–(–S–)$_x$–C covalent bonds that serve to cross-link them into a network structure (see Figure 1.3). You will notice that when you stretch a rubber band a large extension is achieved for a very modest application of force, which, when removed, leads to the complete retraction of the rubber band. This behavior can be contrasted to the response of a steel wire whose cross section is similar to the rubber band. As seen in Figure 1.4, a force that is five orders of magnitude greater than that applied to the rubber band is required to produce a

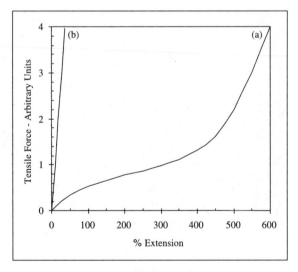

Figure 1.4. Schematic representation of force–extension curve for (a) a rubber-like substance and (b) a monomeric solid.

reversible extension of only 1% in the steel wire. This means that the modulus (stress/strain) of a rubber band is more than 100,000 times lower than the modulus of the steel wire (Treloar, 1975).

Why does the polymer network in the rubber band easily and reversibly stretch to large extensions? Because as the rubber band is macroscopically stretched, so too are the c-PIP chains between the sulfur cross-links. The microscopic extension of the cross-linked c-PIP chains accompanying an extensional force is achieved by rotations about their backbone bonds, resulting in more extended conformations (see Figure 1.3). As you stretch the rubber band, notice that its thickness and width both contract as its length is increased. Overall the volumes of the stretched and at-rest rubber bands are closely similar, meaning that the average separation between c-PIP chains in the network remains unchanged upon stretching. In other words, the additional internal degree of freedom available to the c-PIP chains— that is, their ability to change their sizes and shapes by altering their conformations—lies at the heart of the unique response of elastic polymer networks.

POLYMERS AND TIME

As our third demonstration of physical properties unique to polymers we employ poly(vinyl alcohol) (PVOH), $-(-CH_2-\overset{\overset{\displaystyle OH}{|}}{CH}-)_n-$, in the form of a 5 wt% aqueous solution. Note the thick, viscous, corn syrup-like consistency of the PVOH solution. With stirring, gradually add a few drops of a 5 wt% aqueous solution of borax, $Na_2B_4O_7 \cdot 10H_2O$, which produces a remarkable transformation of the PVOH solution into a gel. Figure 1.5 illustrates the basis for this gel formation. The borate ions $(B(OH)_4^{2-})$ generated upon dissolution of borax are able to establish tetrafunctional interactions (physical, but not covalent) with the hydroxyl groups on PVOH chains, thereby forming temporary, physical cross-links and generating the gel (i.e., swollen network) structure shown in Figure 1.5 (Casassa, Sarquis, and Van Dyke, 1986). The borate ion cross-links are physical and not covalent and may be disrupted if sufficiently strained.

Take the gel out of the beaker, roll it into a cylinder, and hang a portion of the gelled cylinder over the edge of the desktop. Note the

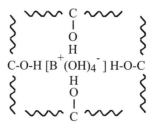

Figure 1.5. Schematic diagram showing the ability of borate ions to form temporary, physical cross-links in PVOH solutions.

slow, irreversible elongation of the cylinder, which eventually extends to the floor without rupture. Next roll the extended gel cylinder into a ball and drop it on the desk surface from a height of 2–3 feet. Notice how the gel ball bounces elastically several times before sticking to and spreading across the desk surface. Finally, roll the gel into another cylinder and attempt to abruptly extend the cylinder by quickly pulling on both of its ends. If pulled quickly enough, the gel cylinder will fail in a brittle manner with very little extension and sharp, flat fracture surfaces.

To summarize, the highly swollen PVOH/borax gel responded in three distinct manners when stressed: (1) It flowed irreversibly when acted upon by gravity as it hung over or rested on the desk, (2) it behaved elastically as it bounced several times on the desktop, and (3) it evidenced brittle failure when rapid extension was attempted. Thus the response of the PVOH gel depended sensitively on the frequency of the deformational force. The temporary borate ion–hydroxyl group, physical cross-links were disrupted and reformed permitting overall irreversible flow of the gel under the steady, low-frequency gravitational force. When the PVOH ball was dropped on the desk, the brief time of impact with the desk was not sufficient to allow the disruption and reformation of the physical cross-links, so the gel ball behaved elastically by bouncing. The attempt to rapidly extend the PVOH gel cylinder led to brittle fracture, because the large extensional force was applied for a time insufficient to permit disruption and reformation of the temporary cross-links and achieve plastic flow or to even allow the PVOH chains between the cross-links to adopt more extended conformations and thereby respond elastically. In each instance the PVOH chains of the temporary, physical network attempted to change their conformations, thereby altering their sizes and shapes in response to stresses of various magnitudes and duration or frequency. Because of their large sizes and temporary connectivity, the PVOH chains are unable to respond instantaneously and instead exhibit frequency- or time-dependent responses. In Chapters 6 and 7 we will revisit the time-dependent macroscopic responses of polymer materials and attempt there to relate them to their microstructurally sensitive internal degrees of freedom, namely their conformations.

POLYMER PHYSICS AND CHEMISTRY

The three macroscopic behaviors that are unique to polymers and were described above—that is, the viscous nature of dilute polymer solutions, the rubber-like elasticity of polymer networks, and the time-dependent responses of swollen polymer gels—do not depend on the *chemistry of polymers* as reflected in their detailed microstructures (see Chapter 4). Instead they are a consequence of the *physics of macromolecules* as dominated by their high-molecular-weight, long-chain natures. Independent of their microstructural constitutions, any polymer may exhibit these three and many other physical behaviors, which are unique to macromolecules. It is only the degrees of their responses, and not their natures, which are influenced by their chemical microstructures. In other words, the physics of long-chain polymers distinguish them from all other classes of materials, while their chemistry only serves to distinguish between the behaviors of different polymers like PEA, PE, PEO, c-PIP, and PVOH, for example.

DISCUSSION QUESTIONS

1. What key structural features distinguish polymers from all other molecules or materials?

2. Select five items that you routinely use and determine whether, how, and why polymers are incorporated into each article.

3. Polyethylene (PE), $-(-CH_2-CH_2-)-_n$, and polystyrene (PS), $-(-CH_2-CH-)-_n$, are commercially important. If a PE with a

molecular weight of 150,000 and a PS with a molecular weight of 550,000 are dissolved in the same solvent at the same concentration, then which solution might be expected to be more viscous, as evidenced by a longer flow time measured with a capillary viscometer (see Figure 1.1)? Why?

4. If the C–C bonds of the PE chain in Question 3 are assumed to adopt three different rotational conformations (*trans*, *gauche+*, *gauche−*; see Figure 1.2), then what is the total number of overall conformations available to this PE?

5. Give a general qualitative description of polymer solids (films, fibers, molded articles, elastomers, etc.) and compare and contrast their physical characteristic/properties to those of a glass tumbler and a steel nail.

6. Based on the brief discussion in this chapter, speculate on possible reasons why the polymeric materials behave differently from the glass tumbler and the steel nail.

7. Use your responses to Questions 5 and 6 to suggest why living things are primarily constucted from the natural polymers proteins, poly(nucleic acids) (DNAs, RNAs), and polysaccharides discussed in Chapter 8, which are often termed *biopolymers*.

REFERENCES

Casassa, E. Z., Sarquis, A. M., and Van Dyke, C. H. (1986), *J. Chem. Educ.*, **63**(1), 57.

Morawetz, H. (1985), *Polymers—The Origins and Growth of a Science*, Wiley-Interscience, New York.

Treloar, L. R. G. (1975), *The Physics of Rubber Elasticity*, 3rd ed., Clarendon Press, Oxford, United Kingdom.

CHAPTER 2

STEP-GROWTH POLYMERIZATION

GENERAL CHARACTERISTICS

Let us suppose that we mix two molecules of the type X–R–X and Y–R'–Y, where X and Y are organic functional groups that react with each other to form a link –Z– (i.e., $-X + Y- \rightarrow -Z-$) and where R and R' represent those portions of both molecules that remain unchanged during the linking reaction. An example of such a linking reaction was mentioned in Chapter 1, where the reaction of ethylene glycol (EG = 1,2-ethanediol), and adipic acid (AA = 1,6-hexanedioic acid), eventually form the polyester, poly(ethylene adipate) (PEA):

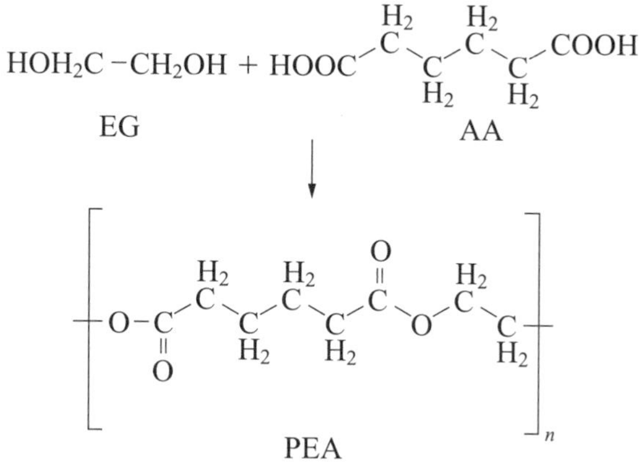

EG   AA

PEA

Returning to our model bifunctional monomers X–R–X and Y–R′–Y, their first reaction on mixing must result in the formation of X–R–Z–R′–Y. Subsequent reaction with either monomer produces X–R–Z–R′–Z–R–X or Y–R′–Z–R–Z–R′–Y. Note that the dimer with one –Z– link and the trimers with two –Z– links are terminated with X,Y; X,X; or Y,Y functional groups and thus may react further with other monomers, with themselves, or with each other. Thus in a stepwise fashion a polymer chain may be grown as indicated schematically below:

$$X-R-X + Y-R'-Y \rightarrow X-R-Z-R'-Y$$

$$X-R-Z-R'-Y + X-R-X \rightarrow X-R-Z-R'-Z-R-X$$

or

$$Y-R'-Y + X-R-Z-R'-Y \rightarrow Y-R'-Z-R-Z-R'-Y$$

$$X-R-Z-R'-Z-R-X + Y-R'-Y \rightarrow X-R-Z-R'-Z-R-Z-R'-Y$$

or

$$X-R-Z-R'-Y + X-R-Z-R'-Y$$
$$\rightarrow X-R-Z-R'-Z-R-Z-R'-Y$$

$$X-R-Z-R'-Z-R-Z-R'-Y + X-R-X$$
$$\rightarrow X-R-Z-R'-Z-R-Z-R'-Z-R-X$$

or

$$X-R-Z-R'-Y + X-R-Z-R'-Z-R-X$$
$$\rightarrow X-R-Z-R'-Z-R-Z-R'-Z-R-X$$

$$X-R-Z-R'Z-R-Z-R'-Z-R-X + Y-R'Y$$
$$\rightarrow X-R-Z-R'-Z-R-Z-R'-Z-R-Z-R'-Y$$

or

$$X-R-Z-R'-Z-R-X + Y-R'-Z-R-Z-R'-Y$$
$$\rightarrow X-R-Z-R'-Z-R-Z-R'-Z-R-Z-R'-Y$$

$$\vdots \qquad\qquad \vdots \qquad\qquad \vdots$$

$$X-(-R-Z-R'-Z-)_m-R'-Y + X-(-R-Z-R'-Z-)_n-R'-Y$$
$$\rightarrow X-(-R-Z-R'-Z-)_{m+n}-R'-Y$$

It should be apparent from the above model reactions that if X,Y = –OH, –COOH and R,R' = –CH$_2$–CH$_2$–, –CH$_2$–CH$_2$–CH$_2$–CH$_2$–, then the polyester PEA would eventually be produced in the step-growth polymerization of EG and AA monomers. The key word in the previous sentence is "eventually," and we will shortly examine its significance. But first consider a single monomer of the type X–R–Y, where again X,Y are organic functional groups that only react to form a link –Z–, –X + Y– → –Z–. Like X–R–X and Y–R'–Y, X–R–Y is bifunctional and, through a stepwise series of reactions similar to those presented above, will eventually produce the polymer X–(–R–Z–)$_n$–R–Y.

By means of the following model experiment, let us examine more closely the step-growth polymerization of a monomer of the type X–R–Y. In a small box place 100 1-inch-square pieces cut from index cards. We assume that each paper square represents an X–R–Y monomer. Their polymerization is begun by reaching into the box and removing two pieces of paper—that is, two monomers—followed by discarding 1 and writing a 2 on the other to denote a dimer and returning it back to the box. This process is repeated and is always accompanied by discarding the drawn piece of paper with the smaller number written on it and then reintroducing into the box the drawn piece of paper with the larger number after it has been increased to the sum of numbers on the two pieces of paper as they were selected from the box. To ensure a random step-growth simulation, the box should be closed and shaken between reaction steps. During the course of this simulated step-growth polymerization, a record of the box contents should be kept, including how many monomers remain, how many oligomers of various lengths have been formed, and the current length of the longest polymer chain that has been formed. In this way we can follow the slow buildup of oligomers to eventually obtain a single polymer chain of 100 repeat units and characterize the reaction products in terms of their average length and the longest chain produced as the conversion of X,Y functional groups proceeds. The results of such a simulated step-growth polymerization are summarized in Table 2.1, where the average length and the length of the longest chain formed are tabulated as a function of the extent of reaction p. The extent of reaction p is simply the fraction of functional groups X,Y that have reacted, that is (the number of links formed)/(the total number of possible links). Note that throughout most of the

TABLE 2.1. Average and Longest Chain Lengths Observed During "in the Box" Simulation of the Chain-Growth Polymerization of X–R–Y

Number of –XY– or –Z– Links Formed*	Longest Chain Formed	Average Chain Length
0	1.0	1.0
1	2.0	1.01
2–14	2.0	1.02–1.16
15–21	3.0	1.18–1.27
22	4.0	1.28
23–45	5.0	1.30–1.82
46–54	6.0	1.85–2.1
55	7.0	2.22
56–58	9.0	2.27–2.38
59	10.0	2.44
60	18.0	2.50
61–76	19.0	2.56–4.17
77–84	21.0	4.35–6.25
85–90	22.0	6.67–10.0
91	28.0	11.1
92	32.0	12.5
93–96	40.0	14.3–25.0
97	55.0	33.3
98	85.0	50.0
99	100	100.0

*p = (number of –XY– or –Z– links)/99

simulated step-growth polymerization both the average and maximum chain size remain very low. Only near the end of the reaction where $p \rightarrow 1.0$ do we begin to obtain long polymer chains. A closer examination of the assumptions built into our simulation of step-growth polymerization is useful for understanding its time evolution and for learning about monomer attributes that are necessary to reach a high extent of reaction p in order to successfully obtain a step-grown polymer with a reasonably high molecular weight.

In our "step-growth polymerization in a box" simulation, where we randomly remove and react monomers, dimers, trimers, and longer oligomers, we have not accounted for any possible differences in the reactivity of functional groups X,Y caused by their attachment to oligomers of different sizes or by the possibility that the reactivity of the X1 group in X1–R–X2 may be affected by whether or not X2 has already reacted. Instead the premise of equal reactivity of functional groups, first put forward and tested by Paul Flory (1953), was assumed in our simulation. Later in this chapter we present some of the data used and gathered by Flory to prove the concept of "Equal Reactivity of Functional Groups," which we employed in our model step-growth polymerization. There we also show how this concept makes possible a simple treatment of the kinetics of step-growth polymerizations.

MOLECULAR WEIGHTS OF STEP-GROWN POLYMERS

In our "in the box" simulation, we observed the average chain length to increase very slowly during the majority of the reaction (for $p = 0.0$–0.9), which is pictorially respresented in Figure 2.1 where the data from Table 2.1 are plotted. The average chain length or number average degree of polymerization $\overline{X}_n$ is simply obtained as the ratio of the number of X–R–Y monomers initially present, N_0, to the number of molecules present after the extent of reaction reaches p, or $N_p = N_0(1 - p)$. Thus,

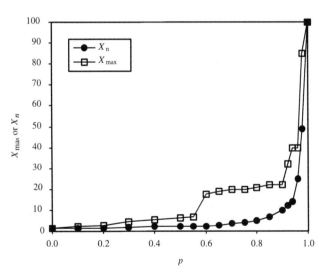

Figure 2.1. Average (●) and longest (□) chain length versus extent of reaction p for "in the box" simulation of step-growth polymerization of 100 X–R–Y monomers.

$$\bar{X}_n = \frac{N_0}{N_0(1-p)} = \frac{1}{(1-p)}$$

which was first presented by Wallace Carothers (Carothers, 1936). This equation is also plotted in Figure 2.1 and is observed to describe our step-growth simulation exactly, as it should. The number average molecular weight is obtained simply from $\bar{M}_n = M_0 . \bar{X}_n$, where M_0 is the molecular weight of the repeat unit $-(-R-Z-)-$.

Because most of the physical properties of polymers show a considerable dependence on their molecular weights, the average molecular weight and the distribution of molecular weights in a polymer sample are both important characteristics of that sample. The Carother's equation $\bar{X}_n = 1/(1-p)$ describes the relationship between the number average degree of polymerization and the extent of reaction in a step-growth polymerization, but because many physical polymer properties are more critically dependent on the largest members belonging to the distribution of polymer chains in a polymer sample, it is very useful to know its weight-average degree of polymerization $\bar{X}_w$ as well, because $\bar{X}_w$ is dominated by the largest polymer chains in the sample. Here we develop expressions for both $\bar{X}_n$ and $\bar{X}_w$ based on a probability approach developed by Flory (1953).

Let us consider a step-growth polymer containing x structural units. This polymer must have been constructed through the reaction of $(x-1)$ functional groups A and has one unreacted A group remaining, where either A–R–A-, B–R'–B-, or A–R–B-type monomers were used. The probability that an A group has reacted is of course p the extent of reaction, so the probability that $(x-1)$A groups have reacted is p^{x-1}, and $(1-p)$ is the probability of finding an unreacted A group. The product of these probabilities is then the probability of finding a polymer with x structural units, $P_x = p^{x-1}(1-p)$. If there are N total polymers in our sample, then clearly the number of polymers containing x structural units is given by $N_x = P_x N = Np^{x-1}(1-p)$. If our step-growth polymerization began with N_0 structural units (monomers) then $N = N_0(1-p)$, and the expression for N_x becomes $N_x = N_0(1-p)^2 p^{x-1}$. Then, $\bar{X}_n$ is given by

$$\bar{X}_n = \frac{\sum_x x N_x}{\sum_x N_X} = \sum_x x P_x$$

So, $\bar{X}_n = \sum_x xp^{x-1}(1-p)$ summed over x, which upon evaluation of the summation yields $\bar{X}_n = 1/(1-p)$, the Carother's equation.

To develop the expression for the weight-average degree of polymerization $\bar{X}_w$, we begin with the weight fraction of polymers in our sample containing x units, W_x. $W_x = xN_x/N_0 = x(1-p)^2 p^{x-1}$. $\bar{X}_w = \sum xW_x = \sum x^2 p^{x-1}(1-p)^2$, which, upon evaluation, becomes $\bar{X}_w = (1+p)/(1-p)$. The breadth of the molecular weight distribution of our step-grown polymer sample may be represented by $\bar{X}_w/\bar{X}_n = (1+p)$ and is usually called the polydispersity index (PDI). PDI $= \bar{X}_w/\bar{X}_n \rightarrow 2.0$ as $p \rightarrow 1.0$ and this has been observed experimentally in a variety of step-growth polymerizations (Odian, 1991).

STEP-GROWTH MONOMERS

At this juncture one might propose the feasibility of making almost a limitless variety of step-growth polymers from all possible monomers of the types X–R–Y or X–R–X and Y–R′–Y, where X,Y can be any pair of reactive organic functional groups. However, this is far from the actual case as can be demonstrated with the Carothers equation and consideration of the equilibria of $-X + Y- \rightarrow -Z-$ linking reactions. There are two general types of linking reactions: $-X + Y- \rightarrow -Z-$ and $-X + Y- \rightarrow -Z- + W$, where W is a new molecule produced and split off during the link formation. Below are examples of both type of linking reactions:

$$-OH + HO-\overset{\overset{\displaystyle O}{\|}}{C}- \longrightarrow -O-\overset{\overset{\displaystyle O}{\|}}{C}- + H_2O$$

$$-OH + O=C=N- \longrightarrow -O-\overset{\overset{\displaystyle O}{\|}}{C}-\underset{H}{N}-$$

where in the formation of the ester bond link a water molecule is produced, but formation of the urethane link from reaction of the hydroxyl and isocyanate groups proceeds without the net production of any additional molecules. Let us now examine the equilibrium aspects of both types of linking reactions.

For $-X + Y- \leftrightarrow -Z-$, the equilibrium constant is $K = [-Z-]/([-X][Y-])$ and provides a measure of the relative conversion of reactants X,Y to products Z. K can be expressed as $p[-X$ or $Y-]_0/\{(1-p)[-X]_0*(1-p)[Y-]_0\}$. For the step-growth polymerization of X–R–Y, $[-X]_0 = [Y-]_0$, so $K = p/\{(1-p)^2[-X$ or $Y-]_0\}$. In order to obtain a reasonably high molecular weight, say $\bar{X}_n = 100$, the Carothers equation requires that $100 = 1/(1-p)$, and so $p = 0.99$. For $p = 0.99$, we see that $K = p/(1-p)^2[-X$ or $Y-]_0 = 0.99/(1-0.99)^2[-X$ or $Y-]_0 = 9900/[-X$ or $Y-]_0$. The initial concentrations $[-X$ or $Y-]_0$ cannot generally exceed 5 moles/liter; therefore, $K > 2000$. The analogous expression for the equilibrium constant describing the linking reaction $-X + Y- \rightarrow -Z- + W$ is easily shown to be $K = p^2/(1-p)^2$, or $K = 9801$ for $p = 0.99$, which is required to achieve $\bar{X}_n = 100$. So we can clearly see that very large equilibrium constants are required for linking reactions that lead to step-growth polymers. In Table 2.2 the five linking reactions with equilibrium constants sufficiently large, or whose equilibria can be driven sufficiently far toward the production of product links by removal of the coproduct split off during their formation, to yield step-growth polymers, are described, and their resultant polymers are introduced briefly below.

STEP-GROWTH POLYMERS

Polyesters

The reaction of dicarboxylic acids $HO\overset{O}{\overset{\|}{-C}}-R-\overset{O}{\overset{\|}{C}}-OH$ with diols HO–R'–OH can produce polyesters, as shown below, where their step-growth linking is through the formation of an ester bond ($-\overset{}{\underset{\|}{\underset{O}{C}}}O-$):

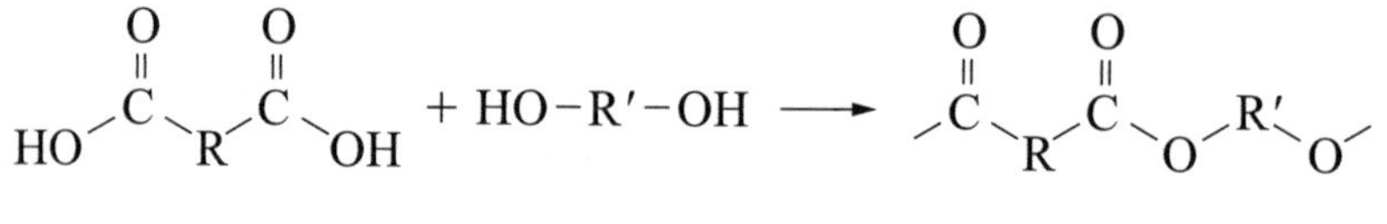

TABLE 2.2. Functional Groups Whose Reaction Lead to Step-Growth Polymers

R–OH Hydroxyl or Alcohol	+	$\overset{\displaystyle O}{\overset{\|}{HO-C-R}}$ Carboxylic acid	→	$\overset{\displaystyle O}{\overset{\|}{R-O-C-R'}}$ Ester	+	H_2O
R–OH Hydroxyl or Alcohol	+	$\overset{\displaystyle O}{\overset{\|}{Cl-C-R'}}$ Acid chloride	→	$\overset{\displaystyle O}{\overset{\|}{R-O-C-R'}}$ Ester	+	HCl
R–NH$_2$ Amine	+	$\overset{\displaystyle O}{\overset{\|}{HO-C-R'}}$ Carboxylic acid	→	$\overset{\displaystyle O}{\overset{\|}{R-\underset{\underset{H}{\|}}{N}-C-R'}}$ Amide	+	H_2O
RNH$_2$ Amine	+	$\overset{\displaystyle O}{\overset{\|}{Cl-C-R'}}$ Acid chloride	→	$\overset{\displaystyle O}{\overset{\|}{R-\underset{\underset{H}{\|}}{N}-C-R'}}$ Amide	+	HCl
R–OH	+	O=C=N–R' Isocyanate	→	$\overset{\displaystyle O}{\overset{\|}{R-O-C-\underset{\underset{H}{\|}}{N}-R'}}$ Urethane		
R–NH$_2$	+	O=C=N–R' Isocyanate	→	$\overset{\displaystyle O}{\overset{\|}{R-\underset{\underset{H}{\|}}{N}-C-\underset{\underset{H}{\|}}{N}-R'}}$ Urea		

Note: $\overset{\displaystyle O}{\overset{\|}{H-C-H}}$ reacts to form –CH$_2$– links with phenol, urea, and melamine, leading to cross-linked, thermosetting, network polymers.

Terephthalic acid and ethylene glycol produce poly(ethylene terephthalate)(PET), which dominates production of synthetic polyesters. PET is principally employed as fibers in apparel and industrial yarns and as blow-molded films used to make beverage containers (Kroschwitz, 1990).

Polyamides

The reaction of dicarboxylic acids HOOC–R–COOH or diacid chlorides ClOC–R–COCl with diamines H_2N–R′–NH_2 can produce polyamides –(–OC–R–CO–NH–R′–NH–)$_n$ by their step-growth linking with –NH–CO– amide bonds. Adipic acid and 1,2-diamino-hexane produce the polyamide, called nylon-6,6.

Nylon-6 [i.e., –(CO–CH_2–CH_2–CH_2–CH_2–CH_2–NH–)$_n$ produced through polymerization of the cyclic amide, caprolactam] and nylon-6,6 are the most widely used polyamides and find extensive applications as fibers in apparel and industrial yarns and in molded applications, such as self-lubricating gears and thick-walled casings for electronic products (Kroschwitz, 1990).

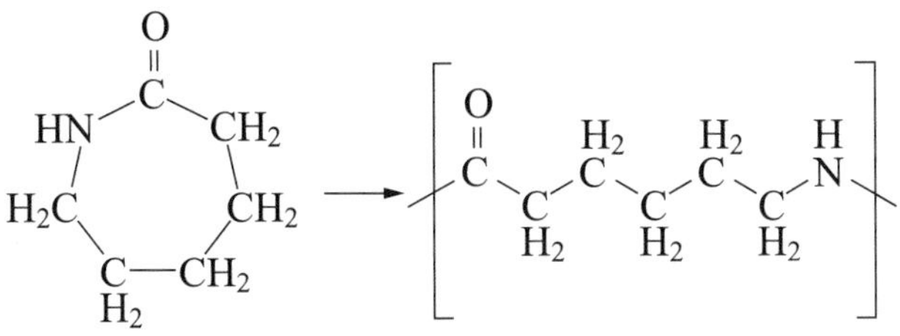

Nylon-6

More recently, several aromatic polyamides have been commercialized, because of their superior strengths and/or thermal stabilities. Poly(*m*-phenylene isothalamide), which is made from the aromatic *m*-diamine and *m*-diacid chloride, is known commercially as Nomex. Nomex is most often used in applications requiring high thermal stability, such as protective clothing for fireman. If the *p*-diamines and *p*-diacid chlorides are employed, the aromatic polyamide, called Kevlar, is produced. Very high strength Kevlar fibers have been produced and are successfully used to make high-performance tire cord and for lightweight armor. Though many times more expensive than their aliphatic counterparts, the polyaramids Nomex and Kevlar possess superior flame and heat resistance, dimensional stability, strength, and modulus (Kroschwitz, 1990; Black and Preston, 1973).

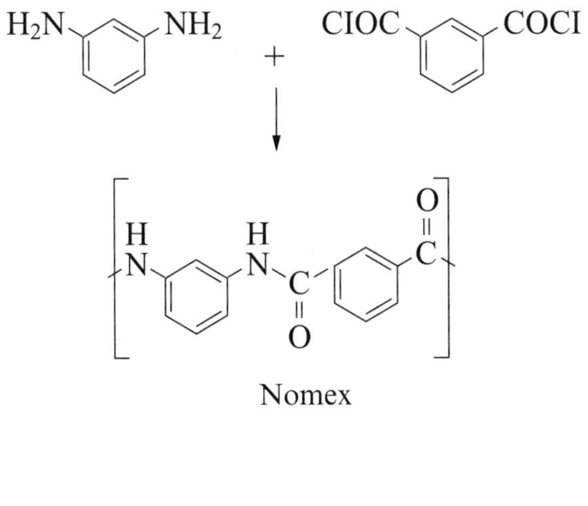

Nomex

Polyurethanes and Polyureas

Reaction of diisocyanates $O=C=N-R-N=C=O$ with diols $HO-R'-OH$ or diamines $H_2N-R'-NH_2$ can lead to polyurethanes

$$\left[R'-O-\overset{\overset{\textstyle O}{\|}}{C}-\underset{\underset{\textstyle H}{|}}{N}-R-\underset{\underset{\textstyle H}{|}}{N}-\overset{\overset{\textstyle O}{\|}}{C}-O \right]_n$$

or polyureas, through the formation of urethane $-\underset{\underset{\textstyle H}{|}}{N}-\overset{\overset{\textstyle O}{\|}}{C}-O-$ and urea

$-\underset{\underset{\textstyle H}{|}}{N}-\overset{\overset{\textstyle O}{\|}}{C}-\underset{\underset{\textstyle H}{|}}{N}-$ linkages:

$$\left[R'-\underset{\underset{\textstyle H}{|}}{N}-\overset{\overset{\textstyle O}{\|}}{C}-\underset{\underset{\textstyle H}{|}}{N}-R-\underset{\underset{\textstyle H}{|}}{N}-\overset{\overset{\textstyle O}{\|}}{C}-\underset{\underset{\textstyle H}{|}}{N} \right]_n$$

However, because both the urethane and urea linkages can add to isocyanate groups, branched and cross-linked structures may also be formed. In addition, isocyanates react with water to form urea linkages,

$$2-N=C=O + H_2O \longrightarrow -\underset{\underset{\textstyle H}{|}}{N}-\overset{\overset{\textstyle O}{\|}}{C}-\underset{\underset{\textstyle H}{|}}{N}- + CO_2$$

with evolution of CO_2, leading to polyurethane foams. Through control of the amounts of diols and diamines in the reaction mixture with diisocyanates and the reaction conditions (temperature and moisture), it is posible to obtain a wide-range of polyurethane/polyurea products that effectively function as flexible and rigid foams, solid elastomers, extrudates, coatings, and adhesives (Kroschwitz, 1990; Woods, 1987).

Formaldehyde Thermosets

Thermosetting, cross-linked polymers may be obtained by linking phenol molecules with formaldehyde, resulting in phenolic resins, or by linking the amino compounds urea or melamine with formaldehyde to yield amino plastics. As illustrated in Figure 2.2, formaldehyde serves to provide methylene links $-CH_2-$ between the trifunctional phenol, tetrafunctional urea, and hexafunctional melamine molecules, which lead eventually to heavily cross-linked thermosetting networks. The phenolic thermosets were the first commercial plastics and were introduced under the trade name Bakelite in 1909 by Henry Baekeland. They remain the largest-volume thermosetting plastic and are characterized by high strength, dimensional stability, and good resistance to impact, creep, solvents, and moisture. The amino plastic thermosets obtained with urea and melamine have properties similar to those of the phenolic thermosets, but are clearer and colorless (Kroschwits, 1990; Odian, 1991; Knop and Pilato, 1985; Vale and Taylor, 1964).

The phenol and amino plastics obtained when formaldehyde reacts with phenol, urea, or melamine are termed *thermosets*, because they are heavily cross-linked, once their networks are formed, they cannot be reshaped by heating or dissolution. This behavior stands in contrast to that exhibited by un-cross-linked polymers, such as PET or nylon-6,6, which may be melted and dissolved after polymerization to alter their macroscopic sample shapes and are termed *thermoplastics*. Thus cross-linked, thermosetting polymers must be polymerized in molds corresponding to the shapes required for their particular applications.

In our remaining discussion of step-growth polymers we focus on the requirements of their linking reactions which must be met in order to achieve a respectible degree of polymerization and consequent molecular weight.

STEP-GROWTH POLYMERIZATION IN AN OPEN SYSTEM

Let us consider the step-growth polymerization of monomers X–R–X and Y–R′–Y, whose functional groups X and Y react according to $-X + Y- \rightarrow -Z- + W$. The equilibrium constant for this reaction is

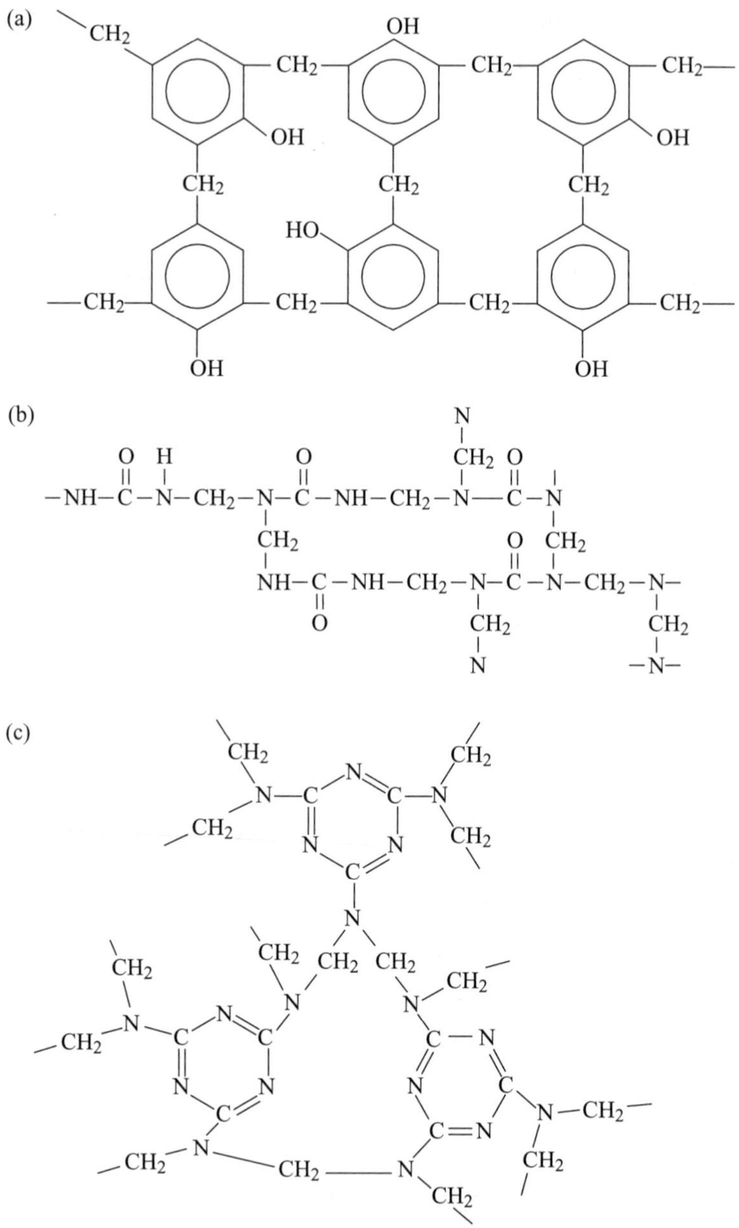

Figure 2.2. Representative structures of (a) phenol-formaldehyde, (b) urea-formaldehyde, and (c) melamine-formaldehyde thermosets.

$K = p[-X \text{ or } Y-]_0[W]\}/\{(1-p)[-X]_0(1-p)[Y-]_0\}$, which reduces to $K = p[W]/\{[-X \text{ or } Y-]_0(1-p)^2\}$. Using $\bar{X}_n = 1/(1-p)$, we obtain $K = p[W]\bar{X}_n^2/[-X \text{ or } Y-]_0$ and $\bar{X}_n(\bar{X}_n - 1) = K[-X \text{ or } Y-]_0/[W]$. Clearly the degree of polymerization $\bar{X}_n$ can be increased by removing the condensation byproduct W produced during the linking reaction. This is illustrated in Table 2.3 for various values of the equilibrium constant K and assuming that the reaction is begun with pure monomers (no solvent), where $[-X]_0 = [Y-]_0 = 5$ M. Remember that K is the ratio of product concentrations to reactant concentrations acheivable in a chemical reaction and is a constant at a given temperature and pressure, independent of these concentrations. Thus for any particular reactant concentrations, the concentrations of the resulting products may be controlled by either removing or adding one of the products to the reaction mixture. In this manner the concentration of the desired or unwanted product may be enhanced or reduced, respectively.

Because the equilibrium constants for polyesterification are typically $K = 0.1–1$, while for polyamidification $K > 100$, removing the water of condensation ($-COOH + HO- \rightarrow -COO + H_2O$) is much more critical in obtaining polyesters with $\bar{X}_n \sim 100$ than for production of reasonably high-molecular-weight nylons. In addition, once the equilibrium constant has been established for a particular step-growth polymerization, the molecular weight or $\bar{X}_n$ of the resulting polymer may be controlled by adjusting the reaction conditions to achieve the appropriate level of remaining byproduct as indicated in Table 2.3 (Odian, 1991).

STOICHIOMETRIC CONTROL OF STEP-GROWTH POLYMERIZATION

In our discussion of step-growth polymerizations we have demonstrated that high-molecular-weight polymers are only achieved near the very end of the reaction of all functional groups on the constituent monomers. This behavior is embodied in the Carothers equation $\bar{X}_n = 1/(1-p)$, where p is the fractional extent of reaction of the functional groups and assumes that the initial concentrations $[-X]_0$ and $[Y-]_0$ are equal. If, however, we have a nonstoichiometric mix-

RENNER LEARNING RESOURCE CENTER
ELGIN COMMUNITY COLLEGE
ELGIN, ILLINOIS 60123

TABLE 2.3. Effect of Water Concentration on Degree of Polymerization in Open, Driven Systems (Odian, 1991)

K	$\overline{X}_n$	$[H_2O]^a$ (moles/liter)
0.1	1.32^b	1.18^b
	20	1.32×10^{-3}
	50	2.04×10^{-4}
	100	5.05×10^{-5}
	200	1.26×10^{-5}
	500	2.00×10^{-6}
1	2^b	2.50^b
	20	1.32×10^{-2}
	50	2.04×10^{-3}
	100	5.05×10^{-4}
	200	1.26×10^{-4}
	500	2.00×10^{-5}
16	5^b	4.00^b
	20	0.211
	50	3.27×1^{-2}
	100	8.10×10^{-3}
	200	2.01×10^{-3}
	500	3.21×10^{-4}
81	10^b	4.50^b
	20	1.07
	50	0.166
	100	4.09×10^{-2}
	200	1.02×10^{-2}
	500	1.63×10^{-3}
361	20^b	4.75^b
	50	0.735
	100	0.183
	200	4.54×10^{-2}
	500	7.25×10^{-3}

^{a}[H_2O] values are for $[M]_0 = 5$.
^{b}These values are for a closed reaction system at equilibrium.

ture of monomers (i.e., $[-X]_0 \neq [Y-]_0$), then the molecular weight achieved by the step-growth polymerization of X–R–X and Y–R′–Y will be reduced from that expected according to the Carothers equation.

Consider the polymerization of X–R–X and Y–R'–Y, where Y–R'–Y is in excess. The numbers of X and Y functional groups N_x and N_y are equal to twice the number of X–R–X and Y–R'–Y molecules, so the stoichiometric imbalance r of the two functional groups is given by $r = (N_x/N_y) \leq 1$. The total number of monomers is simply $(N_x + N_y)/2 = N_x(1 + 1/r)/2$. The extent of reaction p is defined as the fraction of X functional groups reacted, with $(1 - p)X$ groups remaining. Of course, rp and $(1 - rp)$ are the fractions of Y functional groups that are reacted and remain, respectively. $N_x(1 - p)$ and $N_y(1 - rp)$ are the numbers of remaining X and Y functional groups, and their sum is the number of polymer chain ends. Because each polymer has two chain ends, the number of polymers present is given by $[N_x(1 - p) + N_y(1 - rp)]/2$. The number-average degree of polymerization $\bar{X}_n$ is the total number of X–R–X and Y–R'–Y initially present, $N_x(1 + 1/r)/2$, divided by the total number of polymers, $[N_x(1 - p) + N_y(1 - rp)]/2$, so $\bar{X}_n = [N_x(1 + 1/r)/2]/\{[N_x(1 - p) + N_y(1 - rp)]/2\} = \mathbf{(1 + r)/(1 + r - 2rp)}$ (Flory, 1953). For the case of stoichiometric balance, where $r = 1$, we see that $\bar{X}_n$ reduces to $1/(1 - p)$ or the Carothers equation. Although never attainable in practice, when $p = 1$, $\bar{X}_n = (1 + r)/(1 - r)$.

Examination of Figure 2.3 illustrates how the molecular weight of a step-growth polymer may be controlled by adjustment of the stoichiometric imbalance of monomers r. For high extents of reaction p, as little as 1 mol% excess of Y–R'–Y can substantially reduce the molecular weight of the resultant polymers. It is clear that a step-growth polymerization must be carried out to at least $p = 0.98$ if $\bar{X}_n = 50$–100 is to be achieved, which is usually required for a useful polymer. Stoichiometric control of the molecular weight of the step-growth polymerization of X–R–X and Y–R'–Y can also be achieved by addition of a small amount of a monofunctional reactant such as R''–Y to an equimolar mixture of the difunctional monomers. In this instance, $r = N_x/(N_y + 2N_{y'})$, where $N_{y'}$ is the number of mono-functional molecules R''–Y added and $N_x = N_y$. The coefficient 2 is necessary, because one R''–Y molecule has the same effect as one excess Y–R'–Y molecule in limiting the growth of the polymer chain. Another application of controlling the molecular weight of a step-grown polymer by addition of a monofunctional species occurs with X–R–Y-type monomers such as α,ω-hydroxy or amino acids. Clearly $N_x = N_y$, so $r = 1$ in the absence of any added R''–Y or R''–X.

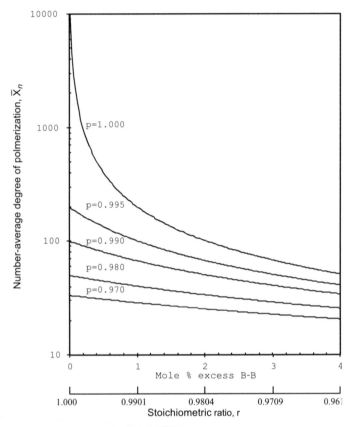

Figure 2.3. Dependence of the number-average degree polymerization $\overline{X}_n$ on the stoichiometric ratio r for different extents of reaction p in the polymerization of A–A with B–B.

Because the ends of these polymers must be terminated by one X and one Y functional group, they are unstable toward changes in molecular weights. Addition of small amounts of R″–X or R″–Y can stabilize their molecular weights, and here, once again, $r = N_x/(N_x + 2N_{y'})$, while $N_x = N_y =$ the number of X–R–Y molecules.

The discussion concerning the control of molecular weight in step-growth polymerizations via functional group stoichiometry presupposes that r remains constant throughout the entire course of polymerization. If one of the monomers were to preferentially volatilize or be removed *via* a side reaction, then the initial stoichiometric imbalance r would tend to increase with the extent of reaction p and the resulting molecular weight would be lowered below that expected based on the initial imbalance. By a judicious continuous or batch-wise addition of the "lost" reactant, it may be possible to maintain the appropriate stoichiometric ratio throughout such a step-growth polymerization.

MOLECULAR WEIGHTS AND THEIR DISTRIBUTION

We have seen that the number-average degree of polymerization achieved in a step-growth polymerization is given by the Carothers equation $\bar{X}_n = 1/(1 - p)$. The number-average molecular weight is of course $\bar{M}_n = \bar{X}_n M_0$, and M_0 is the molecular weight of a repeat unit. For any polymer sample, step-growth or not, its number-average molecular weight is simply

$$\bar{M}_n = \sum_x N_x M_x$$

where N_x and M_x are the mole fractions and molecular weights, respectively, of polymer chains in the sample containing x repeat units. The weight-average molecular weight $\bar{M}_w$ of the same polymer sample is simply

$$\bar{M}_w = \sum_x W_x M_x$$

where W_x and M_x are the weight fractions and molecular weights, respectively, of polymer chains in the sample containing x repeat units. Of course

$$W_x = \frac{N_x M_x}{\sum_x N_x M_x}$$

so

$$\bar{M}_w = \frac{\sum_x N_x M_x^2}{\sum_x N_x M_x}$$

Clearly, for a polymer sample with a distribution of molecular weights, $\overline{M}_w > \overline{M}_n$ because of the M_x^2 weighting of component molecular weights in $\overline{M}_w$. The polydispersity index PDI is often used to characterize the distribution of molecular weights in a polymer sample and is given by PDI $= \overline{M}_w/\overline{M}_n$. In step-growth polymerizations it was demonstrated that $\overline{M}_w = M_0(1+p)/(1-p)$, and so PDI $= \overline{M}_w/\overline{M}_n = [(1+p)/(1-p)]/[1/(1-p)] = 1+p$. As a step-growth polymerization is driven toward completion and $p \rightarrow 1$, we have PDI $\rightarrow 2$.

During the course of a step-growth polymerization of bifunctional monomers X–R–Y or X–R–X and Y–R′–Y, oligomers like X–R–(YX)–R–Y and X–R–(XY)–R′–Y are inevitably produced. If R and R′ are chemical fragments of sufficient conformational flexibilty, such as $-(-CH_2-)_x-$, then there is a small, but not negligible, likelihood that the X,Y functional groups terminating these oligomers can find each other and react to form a cyclic oligomer, which cannot react further. Several mol% of these cycles are often found (Semlyen, 1986) in step-growth polymers like nylons-6 and -6,6 and PET.

KINETICS OF STEP-GROWTH POLYMERIZATION

In order to treat the kinetics of step-growth polymerizations, we must assume that the reaction between two functional groups –X and Y– to form a link –Z– proceeds at the same rate independent of the sizes of the molecules to which they are attached and, for monomers of the type X_1–R–X_2 and Y_1–R′–Y_2, independent of whether or not X_1,X_2 or Y_1,Y_2 have already reacted. As noted by Flory (1953), without this assumption each step in the growth of the polymer would require a different rate constant resulting in a hopelessly complicated kinetic scheme. However, as indicated by the kinetic data in Table 2.4, the concept of equal reactivity of functional groups introduced by Flory is not only a practical necessity to treat the kinetics of step-growth polymerizations, but is in many cases an experimental reality. For the HCl-catalyzed esterification of aliphatic mono- and dicarboxylic acids (R–COOH and HOOC–R–COOH) with ethanol, we see that once R exceeds 2 or 3 carbon atoms in length, ester bonds are produced at the same rate.

Consider the step-growth polymerization of a diol (HO–R–OH) and a diacid (HOOC–R′–COOH) to yield a polyester. Because the reactivity of the hydroxyl –OH and carboxyl –COOH groups toward each other is unaffected by the size of the molecule to which they are

TABLE 2.4. Rate Constants for Esterification (25°C) in Homologous Compounds[a,b]

Molecular Size (x)	$K \times 10^4$ for $H(CH_2)_xCO_2H$	$K \times 10^4$ for $(CH_2)_x(CO_2H)_2$
1	22.1	
2	15.3	6.0
3	7.5	8.7
4	7.5	8.4
5	7.4	7.8
6		7.3
8	7.5	
9	7.4	
11	7.6	
13	7.5	
15	7.7	
17	7.7	

[a] Rate constants are in units of liters/mole-sec.
[b] Data from Bhide and Sudborough (1925).

attached (equal reactivity of functional groups), we need only concern ourselves with the kinetics of simple esterification, which is known to be an acid-catalyzed reaction (Otton and Ratton, 1988). Esterification begins with the protonation of the carboxylic acid,

$$
\underset{\sim\sim\sim\sim\sim\sim}{\overset{O}{\overset{\|}{C}}}\text{-OH} + \text{HA} \underset{k_2}{\overset{k_1}{\rightleftarrows}} \underset{\sim\sim\sim\sim\sim\sim}{\overset{OH}{\overset{|}{C}}}\text{-OH (A}^-)\\ +\text{I}
$$

followed by the reaction of the protonated species I with the alcohol to yield the ester link. HA is the acid catalyst, and it is not consumed during the esterification reaction.

$$
\underset{\sim\sim\sim\sim\sim}{\overset{OH}{\overset{|}{C}}}\text{-OH} + \text{HO}\sim\sim\sim\sim\\ +(\text{A}^-)
$$

$$
\overset{k_3}{\underset{k_4}{\rightleftarrows}} \underset{\underset{+(\text{A}^-)}{\underset{\text{II}}{\sim\sim\sim\sim\sim\text{OH}}}}{\overset{OH}{\overset{|}{\underset{|}{\sim\sim\sim\sim\sim\sim\sim}C}\text{-OH}}} \overset{k_5}{\longrightarrow} \sim\sim\sim\sim\overset{O}{\overset{\|}{C}}\text{-O}\sim\sim\sim\sim+H_2O+HA
$$

Water is continually removed in the polyesterification, so that each of the above reactions are driven to the right in order to reach an extent of reaction ($p > 0.98$) sufficient to produce a reasonably high molecular weight. The rate of polymerization, R_p, can be expressed as the rate of disappearance of the reactant carboxyl groups $-d[-COOH]/dt$, which can be followed experimentally by titration for unreacted carboxyl groups with a base. Because k_4 is vanishingly small when the water of condensation is driven off, and because k_1, k_2, and k_5 are large compared with k_3, R_p is synonomous with the creation of species II; that is, $R_p = -d[COOH]/dt = k_3[C^+(OH)_2][OH]$, where $[COOH]$, $[OH]$, and $[C^+(OH)_2]$ are the concentrations of reactant carboxyl, hydroxyl, and protonated carboxyl(I) groups, respectively.

The concentration of protonated carboxyl groups $[C^+(OH)_2]$ is not easily measured, so we use the equilibrium expression for protonation to eliminate $[C^+(OH)_2]$ in the rate expression for R_p; that is, $K = k_1/k_2 = [C^+(OH)_2]/[COOH][HA]$ or $[C^+(OH)_2] = K[COOH][HA]$ and $R_p = -d[COOH]/dt = k_3K[COOH][OH][HA]$. Two quite different kinetic behaviors arise depending on whether or not a strong acid such as sulfuric acid is added as an external catalyst. If no external acid catalyst is added, then $[HA] = [COOH]$ and the rate expression becomes $R_p = -d[COOH]/dt = k[COOH]^2[OH]$,   where $k = k_3K$. This rate expression is third-order overall, with respect to the reactant concentrations, and second-order in the carboxylic acid concentration, because the carboxylic acid must function both as a reactant and as a catalyst. Because the reactant functional group concentrations must be very nearly stoichiometric, the rate expression for polyesterification becomes $-d[M]/dt = k[M]^3$, where $[M] = [COOH] = [OH]$, or $-d[M]/[M]^3 = kdt$. Upon integration we obtain $2kt = \{1/[M]^2\} - \{[1/[M]_0^2\}$. $[M] = (1 - p)[M]_0$, so in terms of the extent of reaction p we see that $1/(1 - p)^2 = 2[M]_0kt + 1$. When $1/(1 - p)^2$ has been plotted versus t, generally good agreement is observed with polyesterification data as seen in Figure 2.4, where the experimental data are reasonably linear.

When a small amount of strong acid is externally added as a catalyst in polyesterification, $[HA]$ remains constant, and so the rate expression becomes $-d[M]/dt = k'[M]^2$, where $k' = k_3K[HA]$. Integration gives $k't = \{1/[M]\} - \{1/[M]_0\}$, or, in terms of p, $[M]_0k't = \{1/(1 - p)\} - 1$. Because $1/(1 - p) = \bar{X}_n$, $\bar{X}_n = 1 + [M]_0k't$. Clearly a plot of $1/(1 - p)$ versus t should be linear when polyesterification is

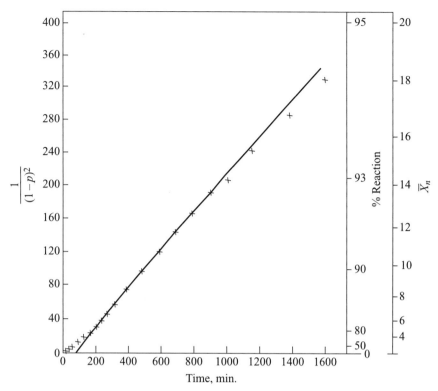

Figure 2.4. Third-order plot of the self-catalyzed polyesterification of adipic acid with diethylene glycol at 166°C. After Solomon (1967) (by permission of Marcel Dekker, Inc., New York) from the data of Flory (1939) (by permission of American Chemical Society, Washington, D.C.).

catalyzed by externally added acid. The same polyesterification represented by the data in Figure 2.4, when catalyzed by the addition of 0.4 mol% p-toluenesulfonic acid, yields the kinetic data shown in Figure 2.5. Clearly, $\bar{X}_n = 1/(1 - p)$ is linear with time, and the time required to achieve a reasonable degree of polymerization $\bar{X}_n$ is greatly reduced compared with the self-catalyzed polyesterification. For this reason, polyesterifications conducted commercially are invariably catalyzed with strong acids. Polyamides can be formed without catalysts; the formaldehyde thermosets formed with phenol, urea, and melamine require acid or base catalysts; and polyurethanes may be produced either with or without employing a catalyst.

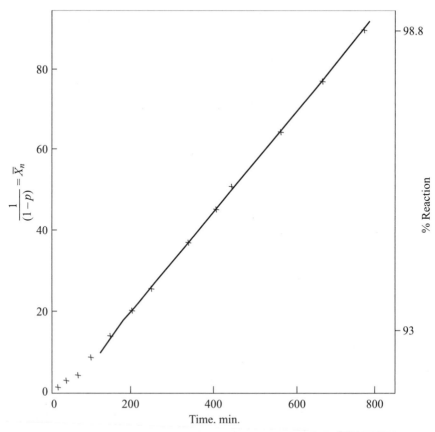

Figure 2.5. Polyesterification of adipic acid with diethylene glycol at 109°C catalyzed by 0.4 mol% *p*-toluenesulfonic acid. After Solomon (1967) (by permission of Marcel Dekker, Inc., New York) from the data of Flory (1939) (by permission of American Chemical Society, Washington, D.C.).

DISCUSSION QUESTIONS

1. From the very large number of reactive functional groups characteristic of all of Organic Chemistry, why are chemists limited to using monomers containing the following functional groups when synthesizing step-growth polymers: $-OH$, $-\overset{\overset{O}{\|}}{C}-OH$, $-NH_2$,

$$\overset{O}{\underset{\|}{}}$$

$-N=C=O$, and $H-\overset{\overset{\displaystyle O}{\|}}{C}-H$ with phenol, urea, or melamine?

2. How can chemists vary the structures of step-growth polymers?

3. Suggest monomers whose step-growth polymerization will lead to linear, branched, and cross-linked polymers. Give examples of each of these polymers.

4. Why must the Carothers equation $\bar{X}_n = 1/(1-p)$ describe the "step-growth polymerization in the box" experiment?

5. Why do you suppose that polyesters and polyamides show different properties even when these monomers HO–R′–OH, $H_2N-R'-NH_2$, and $HO-\overset{\overset{\displaystyle O}{\|}}{C}-R-\overset{\overset{\displaystyle O}{\|}}{C}-OH$ are used in their step-growth polymerizations?

*6. The equilibrium constant $K = 0.8$ for the esterification of $R-\overset{\overset{\displaystyle O}{\|}}{C}-OH + HO-R'$ to produce $R-\overset{\overset{\displaystyle O}{\|}}{C}-O-R' + H_2O$. Using $HO-\overset{\overset{\displaystyle O}{\|}}{C}-R-\overset{\overset{\displaystyle O}{\|}}{C}-OH$ and $HO-R'-OH$ to make the polyester $-(-O-\overset{\overset{\displaystyle O}{\|}}{C}-R-\overset{\overset{\displaystyle O}{\|}}{C}-O-R'-)_n-$, what would be the expected degree of polymerization $\bar{X}_n$ obtained in a closed reaction system and in an open system where the equilibrium H_2O concentration is reduced by a factor of 1000 from its value in the closed system?

*7. The same polyesterification described in Question 6 may be performed in an open system to achieve an extent of reaction $p = 0.99$. What is the corresponding average degree of polymerization $\bar{X}_n$ and how much would it change if 100 moles of the diacid and 101 moles of the diol were used initially?

*8. Compare the times it takes to achieve $p = 0.95$ and $p = 0.99$ for the polyesterification above when conducted in the presence of a strong acid catalyst. If the total time required to achieve $p = 0.99$ is $t(99)$, what fraction of $t(99)$ is required to achieve $p = 0.95$?

REFERENCES

Bhide, B. V., and Sudborough, J. J. (1925), *J. Indian Inst. Sci.*, **8A**, 89.

Black, W. B., and Preston, J., Eds. (1973), *High Modulus Wholly Aromatic Fibers*, Marcel Dekker, New York.

Carothers, W. H. (1936), *Trans. Faraday Soc.*, **32**, 39.

Flory, P. J. (1939), *J. Am. Chem. Soc.*, **61**, 3334.

Flory, P. J. (1953), *Principles of Polymer Chemistry*, Cornell University Press, Ithaca, New York, Chapters III and VIII.

Knop, A., and Pilato, L. (1985), *Phenolic Resins*, Springer-Verlag, Berlin.

Kroschwitz, J. I. (1990), *Concise Encyclopedia of Polymer Science and Engineering*, John Wiley and Sons, New York.

Odian, G. (1991), *Principles of Polymerization*, John Wiley and Sons, New York, Chapter 2.

Otton, J., and Ratton, S. (1988), *J. Polym. Sci., Polym. Chem. Ed.*, **26**, 2183.

Semlyen, J. A., (1986), *Cyclic Polymers*, Elsevier, London.

Solomon, D. H. (1967), *J. Macromol. Sci., Rev. Macromol. Chem.*, **C1**(1), 179.

Vale, C. P., and Taylor, W. G. K. (1964), *Amino Plastics*, Iliffe Books, Ltd., London.

Woods, G. (1987), *The ICI Polyurethane Book*, John Wiley and Sons, New York.

CHAPTER 3

CHAIN-GROWTH POLYMERIZATION

GENERAL CHARACTERISTICS

In addition to step-growth polymerization discussed in the previous chapter, the most frequently used method to synthesize polymers is by the chain-growth polymerization of unsaturated monomers of the type $\overset{R_1}{\underset{R_3}{\diagdown}}C=C\overset{R_2}{\underset{R_4}{\diagup}}$ or $\overset{R}{\underset{\diagup}{\diagdown}}C=C-C=C\overset{\diagup}{\underset{\diagdown}{}}$. The reactivity of the two electrons in the π-bond between sp^2-sp^2, doubly-bonded carbon atoms (see Figure 3.1) lies at the heart of their ability to undergo chain-growth incorporation into long polymer chains. The distant and delocalized nature of the two electrons freely moving in the π-orbital above and below the plane of the trigonally bound sp^2-sp^2 carbon nuclei renders them susceptable to attack by a variety of active species such as free radicals and ions.

Chain-growth polymerization of the vinyl monomer $CH_2=\underset{X}{\overset{|}{C}}H$ must be initiated by a reactive species $R*$, which may be either a free radical or an ion, through opening of the π-bond, addition of the reactive species to the monomer, and formation of a new reactive species as

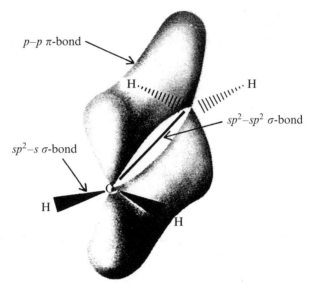

Figure 3.1. The bonding in ethylene consists of one sp^2–sp^2 carbon–carbon σ-bond, four sp^2–s carbon–hydrogen σ-bonds, and one p–p π-bond.

illustrated below:

$$R^* + CH_2{=}\underset{X}{CH} \rightarrow R{-}CH_2{-}\underset{X}{\overset{H}{C^*}}$$

Because $R{-}CH_2{-}\underset{X}{\overset{H}{C^*}}$ is also an active species, the process is repeated as more monomers are successively added and continue to propagate the reactive center. Polymer growth is eventually terminated by destruction of the reactive center through several possible reactions that depend on the type of reactive center and the reaction conditions.

$$R{-}CH_2{-}\underset{X}{\overset{H}{C^*}} \xrightarrow{n\text{-}(CH_2{=}CHX)} R{-}(\!{-}CH_2{-}\underset{X}{\overset{H}{C}}{-}\!)_n{-}CH_2{-}\underset{X}{\overset{H}{C^*}}$$

The chain-growth polymerization of polyacrylamide initiated with riboflavin (vitamin B_2) offers a valuable demonstration of several characteristic aspects of this class of polymerizations (Rodriguez et al., 1987). A 50 wt% aqueous solution of acrylamide monomer is placed in a small beaker, and a few drops of an aqueous solution of riboflavin are added. Nothing happens to the mixed solutions until the beaker is placed on the glass surface of an overhead projector, where it is exposed to UV-VIS radiation of the appropriate wavelengths that cause the riboflavin to photoinitiate free radicals (OH·). As these free radicals initiate several chains, which quickly grow through the rapid addition of acrylamide monomer, the reaction solution begins to sputter and become noticeably warmer with the evolution of steam, because of the highly exothermic nature of the addition of acrylamide monomer to the growing polyacrylamide chains. As the amount of polyacrylamide produced increases, the viscosity of the reaction mixture increases, eventually leading to a leathery disc of water-swollen polyacrylamide containing trapped bubbles of the evolving steam which are clearly visible. The speed and exothermic nature of this chain-growth polymerization are readily apparent, as is the need to expose the riboflavin to light of the appropriate wavelengths in order to produce the requisite initiating free radicals.

CONTRAST BETWEEN CHAIN- AND STEP-GROWTH POLYMERIZATIONS

Chain-growth polymerization of unsaturated monomers obviously proceeds by a mechanism that is clearly distinct from that occurring during the step-growth polymerization of monomers containing at least two functional groups. First the unsaturated monomers must be activated or initiated before they can be incorporated into a polymer chain, while the step-growth monomers begin to react with each other and form links the instant their functional groups come into contact. More important is the difference in the time dependence of polymer production between chain- and step-growth polymerizations. Once the reactive center is produced by the initiator, many monomer units are rapidly added in a chain reaction to produce a high-molecular-weight polymer. Monomer is continually consumed as more chains are initiated and grow rapidly to high molecular weight. The molecular weight of the polymer produced remains relatively unchanged throughout the course of the reaction, even though the overall percent conversion of monomer to polymer increases with the time of reaction. At any time during the course of a chain-growth polymerization the reaction mixture contains monomer, high-molecular-weight polymer, growing chains, and possibly some remaining initiator.

Because, independent of their sizes, any two molecular species can react in a step-growth polymerization, the time dependence of polymer production, as we have seen, is quite different from that observed in a chain-growth polymerization. Monomer disappears much earlier in a step-growth polymerization, as dimers, trimers, tetramers, and so on, are produced. The molecular weight increases only very slowly throughout the course of the step-growth polymerization, and not until the very end of the reaction are high-molecular-weight polymers obtained. Long reaction times are required not only to achieve high percent monomer conversion, but also for obtaining high-molecular-weight polymer. By contrast, in a chain-growth polymerization, high-molecular-weight polymers are generated throughout the course of the reaction with long reaction times only necessary to obtain a high percentage of monomer conversion. High-percentage monomer conversion is critical to achieving high-molecular-weight polymers in a step-growth polymerization, while having little effect on the molecular weight of chain-grown polymers.

It is the distinct natures of step- and chain-growth polymerizations which lead to fundamentally different time dependencies for production of high-molecular-weight polymer, rather than differences in the inherent reactivities of their monomers. This may be dramatically demonstrated by the interfacial polymerization of nylon-6,10 (Morgan and Kwolek, 1959), which has been dubbed "the nylon rope trick" and is presented schematically in Figure 3.2. The two monomers 1,6-diaminohexane and sebacyl chloride are dissolved in separate, immiscible solvents, water, and chlorinated hydrocarbon, respectively. The chlorinated hydrocarbon containing the diacid chloride is denser, and so it is first added to the beaker. Next the less dense, aqueous solution of the diamine is carefully poured on top of the diacid chloride solution to form two distinct phases, which make contact only at their interface. At the interface, the two monomers react to form amide bonds and release HCl which is neutralized by the sodium carbonate present in the upper aqueous phase. A thin film will be

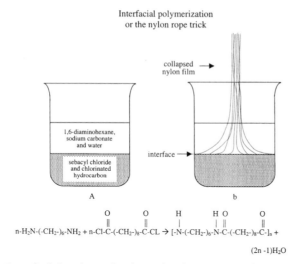

Figure 3.2. Interfacial polymerization of nylon-6,10 or the nylon rope trick (Morgan and Kwolek, 1959).

noticed to form almost instantaneously as the aqueous solution is floated atop the denser chlorinated hydrocarbon phase. This film is nylon-6,10, whose name is derived from the 6 and 10 carbon fragments making up its repeat unit as indicated in Figure 3.2.

If the nylon-6,10 film is raised at its center with tweasers, then a new layer is immediately formed by the rapid polymerization of the two monomers at the interface. In this manner a small rope of nylon-6,10 may be created by slowly winding up the collapsed nylon film (see Figure 3.2). How is it possible to rapidly produce nylon-6,10 in this manner? The key to answering this question lies in the realization that by isolating the two monomers in separate phases we restrict their contact and subsequent reaction to the interface. By this means we have effectively prevented the step growth of nylon-6,10, because the monomers are only able to react with the chain ends of the polymers growing at the interface and not with each other or with various small oligomers as is the case in their homogeneous step-growth polymerization.

Clearly then, long times are required to obtain high-molecular-weight polymers by homogeneous, step-growth polymerization, because the reaction of their monomers must be driven nearly to completion and not because their monomers react slowly.

DETAILS OF CHAIN-GROWTH POLYMERIZATION

Let us examine in greater detail the critical events occurring in a chain-growth polymerization (Odian, 1991). We choose styrene, $CH_2=\overset{\overset{\text{H}}{|}}{C}$, as

our monomer and choose a free radical initiator t-butyl peroxide, $CH_3-\overset{\overset{\text{CH}_3}{|}}{\underset{\underset{\text{CH}_3}{|}}{C}}-O-O-\overset{\overset{\text{CH}_3}{|}}{\underset{\underset{\text{CH}_3}{|}}{C}}-CH_3$, which undergoes thermal, homolytic dissociation to $2CH_3-\overset{\overset{\text{CH}_3}{|}}{\underset{\underset{\text{CH}_3}{|}}{C}}-O^*$ free radicals, each with an unpaired electron, which are necessary to begin our chain-growth polymerization by reaction with a styrene monomer. As a consequence, the initiation step in chain-growth polymerization may be viewed as two consecutive reactions. The first is production of free radicals by the disassociation of our initiator

$$CH_3-\overset{\overset{\text{CH}_3}{|}}{\underset{\underset{\text{CH}_3}{|}}{C}}-O-O-\overset{\overset{\text{CH}_3}{|}}{\underset{\underset{\text{CH}_3}{|}}{C}}-CH_3 \longrightarrow 2CH_3-\overset{\overset{\text{CH}_3}{|}}{\underset{\underset{\text{CH}_3}{|}}{C}}-O^*$$

and the second is the addition of the first monomer to produce the chain initiating species

$$CH_3-\overset{\overset{\text{CH}_3}{|}}{\underset{\underset{\text{CH}_3}{|}}{C}}-O^* + CH_2=\overset{\overset{\text{H}}{|}}{\underset{\underset{\bigcirc}{}}{C}} \longrightarrow CH_3-\overset{\overset{\text{CH}_3}{|}}{\underset{\underset{\text{CH}_3}{|}}{C}}-O-CH_2-\overset{\overset{\text{H}}{|}}{\underset{\underset{\bigcirc}{}}{C}}{}^*$$

Note that the free radical fragment $CH_3-\overset{\overset{\text{CH}_3}{|}}{\underset{\underset{\text{CH}_3}{|}}{C}}-O-$ is chemically bound to the first monomer unit and will remain attached to the polymer both during its rapid growth and after polymerization is complete.

After the initiation step has produced the chain initiating species, monomer is rapidly added and the radical active center is propagated to the end of the growing chain. Each monomer addition simply increases the length of the growing chain without necessarily affecting the reactivity of the radical center, and in combination they are termed the propagation step.

$$
\begin{array}{c}
\underset{\substack{\displaystyle CH_3 \\ \displaystyle | \\ \displaystyle CH_3}}{CH_3-C-O-CH_2-\underset{\bigcirc}{\overset{H}{\underset{|}{C^*}}}} + n(CH_2=\overset{H}{\underset{\bigcirc}{\underset{|}{C}}}) \longrightarrow \longrightarrow \cdots
\end{array}
$$

$$
\longrightarrow CH_3-\overset{CH_3}{\underset{CH_3}{\underset{|}{\overset{|}{C}}}}-O-(-CH_2-\overset{H}{\underset{\bigcirc}{\underset{|}{C}}}-)_n-CH_2-\overset{H}{\underset{\bigcirc}{\underset{|}{C^*}}}
$$

Eventually each of the initiated and growing chains stops growing or is terminated. The termination step annihilates the radical centers on the growing chain by a bimolecular reaction between radicals. Two radicals may react with each other by direct combination or coupling, thereby joining two growing chains into a single larger chain which is "dead," because it cannot further add any monomer, or, more rarely, by disproportionation.

$$
-CH_2-\overset{H}{\underset{\bigcirc}{\underset{|}{C^*}}} + {}^*\overset{H}{\underset{\bigcirc}{\underset{|}{C}}}-CH_2- \longrightarrow -CH_2-\overset{H}{\underset{\bigcirc}{\underset{|}{C}}}\text{———}\overset{H}{\underset{\bigcirc}{\underset{|}{C}}}-CH_2- \quad \text{(Coupling)}
$$

In disproportionation a hydrogen that is beta to a radical center is transferred to another radical center. The resulting "dead" and "dormant" polymer chains created during termination by disproportionation are fully saturated and terminally unsaturated, respectively. The unsaturated chain could conceivably be reinitiated through contact with initiator or the radical terminus of a growing chain.

$$
-CH_2-\overset{H}{\underset{\bigcirc}{\underset{|}{C^*}}} + {}^*\overset{H}{\underset{\bigcirc}{\underset{|}{C}}}-CH_2- \longrightarrow -CH_2-\overset{H}{\underset{\bigcirc}{\underset{|}{C}}}-H + \overset{H}{\underset{\bigcirc}{\underset{|}{C}}}=\overset{H}{\underset{|}{C}}- \quad \text{(Disproportionation)}
$$

CHAIN-GROWN POLYMER SIZES

The sizes of polymer chains produced during a chain-growth poly-merization are determined by the rates of initiation, propagation, and termination. For every radical that initiates the growth of a polymer chain, the average number of monomers incorporated in the growing chain is termed the kinetic chain length v, which is simply given by the ratio of the rate of propagation R_p to the rate of initiation R_i or rate of termination R_t, $v = R_p/R_i = R_p/R_t$, because it is assumed that $R_i = R_t$ (Kondratiev, 1969). The concentration of radicals increases initially, but quickly reaches a constant, "steady-state" value, so $R_i = R_t$.

Any attempt to increase the number of growing chains by increasing the number of radicals produced will decrease the average length of polymers produced, because v is inversely proportional to R_i. The number-average degree of polymerization $\bar{X}_n$ is related to the kinetic chain length v. If a chain is terminated by the coupling of two kinetic chains to form a "dead" polymer, then $\bar{X}_n = 2v$. When a growing chain is terminated by disproportionation, $\bar{X}_n = v$.

Because the mode of termination of chain-growth influences the expected molecular weight, $\bar{X}_n = 2v$ for coupling and $\bar{X}_n = v$ for dis-proportionation, it is important to identify which or how much of each type of termination is occurring in a given chain-growth poly-merization. This can be achieved (Odian, 1991) by determining the number of initiator fragments per polymer molecule b, which is obtained by measuring the molecular weight of the polymer and the number of initiator fragments contained in the polymer sample. If a is taken as the fraction of propagating chains terminated by coupling and $(1 - a)$ is taken as the fraction terminated by disproportionation, then n propagating chains will yield an initiator fragments and $an/2$ polymer molecules for termination by coupling and $(1 - a)n$ initiator fragments and $(1 - a)n$ polymer molecules for termination by dispro-portionation. The average number of initiator fragments per poly-mer is then simply the total number of initiator fragments divided by the total number of polymer chains, so $b = [an + (1 - a)n]/[an/2 + (1 - a)n] = 2/(2 - a)$. Consequently the fractions of chains terminated by coupling and disproportionation are $a = (2b - 2)/b$ and $(1 - a) = (2 - b)/b$, respectively. It has been generally observed that termination of free-radically initiated chain-growth polymers most often occurs by coupling (Tedder, 1974).

TRANSFER OF CHAIN GROWTH

Occasionally a growing chain is prematurely terminated by transfer of a hydrogen or other atom or species from some compound in the system. This compound may be the monomer, initiator, solvent, or the polymer. If M_n^* represents the growing chain and XA represents the agent that transfers the atom or species X, then chain transfer may be depicted as $M_n^* + XA \rightarrow M_n-X + A^*$. The growing chain is terminated and a new radical species A* is produced which may re-initiate polymerization $A^* + M \rightarrow A-M^*$. Chain transfer reduces the size of the propagating polymer chain and may also affect the overall rate of polymerization depending on the relative rates of reinitiation and the usual propagation of the growing chain. In those polymerization systems where the molecular weight is observed to be lower than expected based on the experimentally observed extents of termination by coupling and disproportionation, some type of chain transfer reaction must be occurring.

In addition to possibly lowering the expected molecular weight of a chain-grown polymer, when chain transfer occurs to polymer a new radical site is produced on the interior of the polymer, resulting in the formation of a branch. For example,

$$M_n^* + -CH_2-\overset{\overset{\displaystyle Y}{|}}{\underset{\underset{\displaystyle H}{|}}{C}}- \longrightarrow M_n-H + -CH_2-\overset{\overset{\displaystyle Y}{|}}{\underset{\underset{\displaystyle *}{|}}{C}}- \overset{M_m}{\longrightarrow} -CH_2-\overset{\overset{\displaystyle Y}{|}}{\underset{\underset{\displaystyle *}{\underset{\displaystyle M_m}{|}}}{C}}-$$

Beyond the potential reduction in molecular weight, the production of branches by chain transfer to polymer can seriously affect the processing and physical properties of the polymer. On the other hand, we may control the molecular weight of a chain-grown polymer by purposely adding a chain transfer agent to the reaction mixture.

CHAIN-GROWN POLYMERS

Some common examples of synthetic polymers made by chain-growth polymerization are presented in Table 3.1.

Of course when two different unsaturated vinyl monomers are placed in the reaction vessel along with a free radical initiator, their chain-growth polymerization may result in the incorporation of both monomers in the growing chains to produce copolymers. Depending on the relative reactivities of each of the monomers A and B toward the radical ends –A* and –B* of the growing chains, the resultant copolymers may have a random (–A–B–B–A–B–A–A–A–B–), regularly alternating (–A–B–A–B–A–B–A–B–), or blocky (–A–A–A–A–A–A–B–B–B–B–B–) microstructure. The physical properties of an A–B copolymer may be quite distinct from the constituent homopolymers poly-A and poly-B, and the copolymer properties may also depend sensitively on the distribution of comonomers in the copolymer—that is, whether or not the copolymer is random, regularly alternating, or blocky (see Chapter 7). Clearly the ability to copolymerize two monomers and to control the comonomer sequence in the resulting copolymer gives the polymer chemist a much expanded range of options when synthesizing polymers to achieve properties that satisfy the requirements of specific materials applications.

Our remaining discussion addresses the kinetic aspects of chain-growth homo- and copolymerizations and two variants of chain-growth polymerization termed emulsion and living polymerizations.

KINETICS OF CHAIN-GROWTH POLYMERIZATION

Here we examine the kinetics or rate of free-radical initiated chain-growth polymerization. The initiation step involves two reactions. The first is generation of free radicals R* from the initiator I, that is,

$$I \xrightarrow{k_d} 2R^*$$

The second reaction in the initiation step involves the addition of R* to the first monomer M to produce the chain initiating species R–M* = M1*, that is,

$$R^* + M \xrightarrow{k_i} M1^*$$

TABLE 3.1. Some Commercial Chain-Growth (Vinyl) Polymers Prepared by Free-Radical Polymerization

Monomer	Formula	Polymer	Uses
Ethylene	$CH_2{=}CH_2$	Polyethylene	Sheets and films, blow-molded bottles, injection-molded toys and housewares, wire and cable coverings, shipping containers
Propylene	$CH_2{=}CHCH_3$	Polypropylene	Fiber products such as indoor–outdoor carpeting, car and truck parts, packaging, toys, housewares
Styrene	$CH_2{=}CH-\bigcirc$	Polystyrene	Packaging and containers (Styrofoam), toys, recreational equipment, appliance parts, disposable food containers and utensils, insulation
Acrylonitrile (propenenitrile)	$CH_2{=}CHC{\equiv}N$	Polyacrylonitrile (Orlon, Acrylan)	Sweaters and other clothing
Vinyl acetate (ethenyl ethanoate)	$CH_2{=}CH-O\overset{O}{\overset{\|}{C}}CH_3$	Polyvinyl acetate	Adhesives, latex paints
Methyl methacrylate (methyl 2-methyl-propenoate)	$CH_2{=}C(CH_3)-C\overset{O}{\overset{\|}{}}OCH_3$	Polymethyl methacrylate (Plexiglas, Lucite)	Objects that must be clear, tranparent, and tough
Vinyl chloride (chloroethene)	$CH_2{=}CHCl$	Polyvinyl chloride (PVC)	Plastic pipe and pipe fittings, films and sheets, floor tile, records, coatings
Tetrafluoroethylene (tetrafluoroethene)	$CF_2{=}CF_2$	Polytetrafluoroethylene (Teflon)	Coatings for utensils, electric insulators

k_d and k_i are the rate constants for the above two initiating reactions. Successive additions of monomers to M1* constitute the propagation step, which can be individually expressed as the addition of M to M_n^* to produce M_{n+1}^*,

$$M_n^* + M \xrightarrow{k_p} M_{n+1}^*$$

k_p is the rate constant for propagation. Termination by coupling and disproportionation can be represented by the following expressions:

$$M_n^* + M_m^* \xrightarrow{k_{tc}} M_{n+m}$$
$$M_n^* + M_n^* \xrightarrow{k_{td}} M_n + M_m$$

k_{tc} and k_{td} are the rate constants for termination by coupling and disproportionation, respectively. The termination step can also be represented by

$$M_n^* + M_m^* \xrightarrow{k_t} \text{dead polymer}$$

where $k_t = k_{tc} + k_{td}$.

We can see that monomer is consumed in both the initiation and propagation steps of the chain-growth polymerization, so $-d[M]/dt = R_i + R_p$, where R_i and R_p are the rates of initiation and propagation, respectively. Because the numbers of monomers reacting in the initiation reaction is so much smaller than the number consumed in the propagation steps, to a good approximation $-d[M]/dt = R_p$. Of course $R_p = k[M^*][M]$, where $[M]$ is the monomer concentration and $[M^*]$ is the concentration of all radicals. Radical concentrations are difficult to measure, so we seek to eliminate $[M^*]$ from the rate expression for polymerization. To achieve this elimination we assume that the concentration of radicals almost instantaneously reaches a constant steady value, which means that the rate of change of $[M^*]$ quickly becomes and remains zero during the course of the polymerization. Because M* are created during initiation and consumed during termination, $-d[M^*]/dt = 0$ means that the rates of initiation and termination are equal: $R_i = R_t$. Therefore, $R_t = 2k_t[M^*]^2 = R_i$, and $[M^*] = (R_i/2k_t)^{1/2}$, which may be substituted into the expression for R_p to yield $R_p = k_p[M](R_i/2k_t)^{1/2}$.

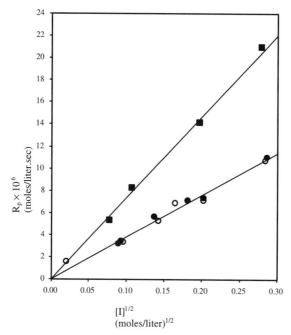

Figure 3.3. Square-root dependence of the polymerization rate R_p on the initiator concentration [I]. ■: Methyl methacrylate, benzoyl peroxide, 50°C. After Schulz and Blaschke (1942) (by permission of Akademische Verlagsgesellschaft, Geest and Portig K.-G., Leipzig). ●, ○: Vinyl benzoate, azobisisobutyronitrile, 60°C. After Santee et al. (1964) and Vrancher and Smets (1959) (by permission of Huthig and Wepf Verlag, Basel).

We have seen that the initiation step in chain-growth polymerizations is composed of two reactions $I \xrightarrow{k_d} 2R^*$ and $R^* + M \xrightarrow{k_i} M1^*$ whose rates are very different. Addition of M to R^* is generally much faster than generation of R^* from I, so we may consider $I \xrightarrow{k_d} 2R^*$ as the rate-determining reaction in the initiation step and write $R_i = 2fk_d[I]$, where f is the initiator efficiency, generally smaller than 1, because not every R^* produced initiates the growth of a polymer chain. If we now substitute this expression for R_i into our expression for the rate of polymerization $\{R_p = k_p[M](R_i/2k_t)^{1/2}\}$, then $R_p = k_p[M]\{(fk_d[I])/k_t\}^{1/2}$.

Figures 3.3 and 3.4 present rate of chain-growth polymerization data illustrating the dependence on both initiator and monomer con-

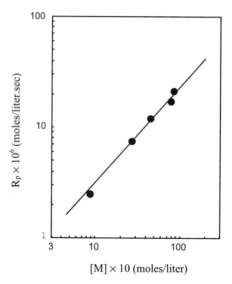

Figure 3.4. First-order dependence of the polymerization rate R_p of methyl methacrylate on the monomer concentration [M]. The initiator is the t-butyl perbenzoate–diphenylthiourea redox system. After Sugimura and Minoura (1966) (by permission of Wiley-Interscience, New York).

centrations, $[I]^{1/2}$ and [M] (Schulz and Blaschke, 1942; Santee et al., 1964; Vrancher and Smets, 1959; Sugimura and Minoura, 1966). Note that R_p depends linearly on both $[I]^{1/2}$ and [M] as predicted by the rate expression $R_p = k_p[M]\{(fk_d[I])/k_t\}^{1/2}$. This behavior is commonly observed in many free-radical-initiated chain-growth polymerizations.

The kinetic chain length $v = R_p/R_i = R_p/R_t$, and, with the previous expressions developed for R_p, R_i, and R_t, becomes $v = (k_p[M])/2(fk_dk_t[I])^{1/2}$. Note that v is inversely proportional to $[I]^{1/2}$, while R_p is directly proportional to $[I]^{1/2}$. This means we can increase the rate of polymerization by increasing the initiator concentration, but only at the expense of reducing the molecular weight of the polymers formed.

The molecular weight of a chain-grown polymer produced under conditions where chain transfer reactions are occurring depends on the rate of polymerization divided by the sum of the rates of all chain transfer reactions plus the rate of chain termination; that is, $\bar{X}_n = R_p/\{(R_t/2) + \Sigma R_{tr}\}$, where ΣR_{tr} are the rates of transfer of the radical center on the growing chain to the monomer, solvent, initiator, or polymer. Because these rates depend on the monomer and initiator concentrations that change during the course of the reaction, the degree of polymerization or the molecular weight of a chain-grown polymer will depend on when it was initiated. At low conversion with termination by disproportional only and with chain transfer, it can be demonstrated (Flory, 1953) that $\bar{X}_n = 1/(1 - p)$, $\bar{X}_w = (1 + p)/(1 - p)$, and PDI $= \bar{X}_w/\bar{X}_n = (1 + p)$ as previously demonstrated for step-growth polymerizations. In the case of chain-growth polymerizations, p is defined as the probability that a propagating radical will continue to propagate rather than terminate, so $p = R_p/(R_p + R_t + \Sigma R_{tr})$.

When termination is by coupling only and no chain transfer is occurring, the molecular weights and their distribution can be shown (Flory, 1953) to be described by $\bar{X}_n = 2/(1 - p)$, $\bar{X}_w = (2 + p)/(1 - p)$, and **PDI** $= \bar{X}_w/\bar{X}_n = (2 + p)/2$.

For high-conversion chain-growth polymerizations the size distributions of chains obtained are much broader than at low conversions or those found in step-growth polymerizations. We saw that the kinetic chain length v depends on $[M]/[I]^{1/2}$. Under most reaction conditions $[I]^{1/2}$ decreases faster than $[M]$, so the molecular weight of the polymer produced at any instant will increase with conversion. Both R_p and v are proportional to $k_p/k_t^{1/2}$. As the conversion increases, producing more and more polymer, the viscosity of the reaction medium increases. Because termination requires the movement of two growing chains, while propagation only requires movement of monomer, the ratio $k_p/k_t^{1/2}$ is expected to increase dramatically, leading to large increases in both the rate of polymerization and the degree of polymerization at high conversions. This behavior is often observed and has been termed *autoacceleration* (North, 1974). If chain transfer to polymer occurs, PDI $= 5$–10; values of 20–50 are commonly observed (Odian, 1991) in chain-grown polymers produced under conditions leading to high conversion.

CHAIN-GROWN COPOLYMERS AND THE COPOLYMER EQUATION

Let us now consider the chain-growth copolymerization of vinyl monomers M1 and M2. During the propagation of the growing copolymer chain the following four reactions play a crucial role in determining the incorporation and therefore the sequence of comonomers M1 and M2 in the final copolymer:

$$-M1^* + M1 \xrightarrow{k_{11}} -M1^*$$

$$-M1^* + M2 \xrightarrow{k_{12}} -M2^*$$

$$-M2^* + M1 \xrightarrow{k_{21}} -M1^*$$

$$-M2^* + M2 \xrightarrow{k_{22}} -M2^*$$

In the following development, we are assuming that the reactivity of a growing chain depends only on the monomer at the growing end, that is, $-M1^*$ or $-M2^*$. This is termed the first-order Markov or terminal model of copolymerization, where k_{11} and k_{12} are the rate constants for adding M1 and M2, respectively, to a chain with a growing $M1^*$ end. Likewise for chains with a growing $M2^*$ end, k_{22} and k_{21} are the rate constants for addition of M2 and M1, respectively.

M1 disappears in the first and third reactions, while M2 is consumed in the second and fourth, and because their rates of disappearance must be synonomous with their rates of incorporation into the copolymer, we may write them as $-d[M1]/dt = k_{11}[M1^*][M1] + k_{21}[M2^*][M1]$ and $-d[M2]/dt = k_{12}[M1^*][M2] + k_{22}[M2^*][M2]$. Dividing the first expression by the second yields the ratio of rates at which the two comonomers are incorporated into the copolymer, that is, $d[M1]/d[M2] = (k_{11}[M1^*][M1] + k_{21}[M2^*]\{M1\})/(k_{12}[M1^*][M2] + k_{22}[M2^*][M2])$. To remove the difficult-to-measure $[M1^*]$ and $[M2^*]$ concentrations from this expression, it is assumed that both concentrations remain constant (steady state), which means that the rates of the second and third reactions, which describe monomer additions that change the nature of the radical center, must be equal, $k_{21}[M2^*][M1] = k_{12}[M1^*][M2]$. Combination of these two immediately preceding equations and letting $r_1 = k_{11}/k_{12}$ and $r_2 = k_{22}/k_{21}$ results in the following expresssion known as the copolymer equation: $d[M1]/d[M2] = \{[M1](r_1[M1] + [M2])\}/\{[M2]([M1] + r_2[M2])\}$. The copolymer equation describes the incorporation of monomers in the copolymer $d[M1]/d[M2]$ in terms of the concentrations of monomers in the feed $[M1]$ and $[M2]$ and the monomer reactivity ratios r_1 and r_2. Each r is the ratio of the rate constant for a propagating species adding its own monomer to the rate constant for adding the other monomer (Alfrey and Goldfinger, 1944; Mayo and Lewis, 1944; Wall, 1944; Walling, 1957).

It must be stressed that the copolymer equation describes the instantaneous composition of comonomers incorporated into the copolymer chain ($d[M1]/d[M2]$) at the same instant that the monomer feed ratio is $[M1]/[M2]$. Of course in most copolymerizations that are begun with a comonomer feed ratio of $([M1]/[M2])_0$ the feed ratio will continue to change as the two monomers are incorporated at different rates. Consequently at time $= t$, $([M1]/[M2])_t \neq ([M1]/[M2])_0$ and so the comonomer compositions incorporated initially and at time t will not be the same, that is, $(d[M1]/d[M2])_0 \neq (d[M1]/d[M2])_t$. For this reason we can expect the comonomer sequence to be continually changing as the feed ratio changes, and certainly those copolymers whose chain growth is initiated early in the reaction will have very different comonomer sequence distributions than those copolymers initiated at higher extents of reaction.

A more convenient form of the copolymer equation results if f_1 and f_2, the mole fractions of M1 and M2 in the feed, $f_1 = 1 - f_2 = [M1]/([M1] + [M2])$, and F_1 and F_2, the mole fractions of M1 and M2 incorporated in the copolymer, $F_1 = 1 - F_2 = d[M1]/(d[M1] + d[M2])$, are used, that is, $\boldsymbol{F_1 = (r_1 f_1^2 + f_1 f_2)/(r_1 f_1^2 + 2 f_1 f_2 + r_2 f_2^2)}$. The copolymerization equation has been shown to be applicable to many copolymerizations (Odian, 1991), including ionically initiated chain-growth copolymerizations. However, for any comonomer pair M1 and M2, r_1 and r_2 can drastically differ depending on the mode of initiation, which once again permits a greater range of potentially attainable copolymer microstructures and concomitant properties (Landler, 1950; Pepper, 1954).

The comonomer sequence distribution attainable in a copolymer is governed by both the feed ratio f_1/f_2 and the reactivity ratios r_1 and r_2. Two monomers will copolymerize when r_1 and r_2 are between zero and unity. For r_1 and r_2 greater than unity means that M1* and M2* preferentially add M1 and M2 instead of M2 and M1, while M1* and M2* preferentially add M2 and M1 when r_1 and r_2 are less than unity. Three types of copolymerization behavior are observed depending on whether $r_1 r_2 = 1$, <1, or >1. The copolymerization is termed *ideal* when $r_1 r_2 = 1$, because in this instance $r_1 = 1/r_2$, or $k_{11}/k_{12} = k_{21}/k_{22}$, which means that both propagating species M1* and M2* show the same preference for adding one or the other of

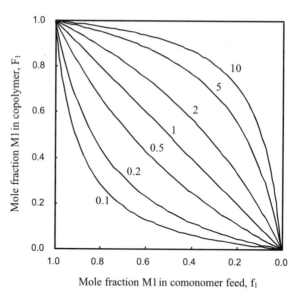

Figure 3.5. Dependence of the instantaneous copolymer composition F_1 on the initial comonomer feed composition f_1 for the indicated values of r_1 where $r_1 r_2 = 1$. After Walling (1957) (by permission of Wiley, New York) from plot in Mayo and Walling (1950) (by permission of American Chemical Society, Washington, D.C.).

the two monomers M1 and M2. For ideal copolymerizations, the copolymer equation reduces to $F_1 = r_1 f_1 / (r_1 f_1 + f_2)$, which is plotted in Figure 3.5. When $r_1 = r_2 = 1$ the two monomers are equally reactive toward both propagating species resulting in $F_1 = f_1$ or a coincidence between the copolymer and monomer feed compositions. The resulting polymers are called random or Bernoullian. For copolymers with $r_1, r_2 < 1$ and $r_1, r_2 > 1$, one of the monomers is more reactive toward both propagating species than is the other monomer. The copolymer will contain a higher proportion of the more reactive monomer in random placement. Ideal copolymers formed from monomers with very disparate r_1 and r_2, such as $r_1 = 10$ and $r_2 = 0.1$, cannot contain appreciable amounts of both monomers. As an example, in the case above with a monomer feed ratio of $f_2 = 0.8$ (80 mol% M2), only 18.5 mol% M2 ($F_2 = 0.185$) is incorporated in the copolymer. Significant incorporation of both monomers only occurs when r_1 and r_2 do not differ markedly ($r_{1,2} = 0.5$–2, for example).

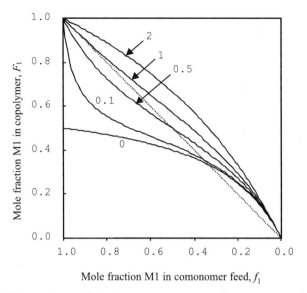

Figure 3.6. Dependence of the instantaneous copolymer composition F_1 on the initial comonomer feed composition f_1 for the indicated values of r_1 with r_2 being constant at 0.5. After Walling (1957) (by permission of Wiley, New York) from plot in Mayo and Walling (1950) (by permission of American Chemical Society, Washington, D.C.).

When $r_1 = r_2 = 0$, the copolymer equation becomes $d[M1]/d[M2] = 1$ or $F_1 = 0.5$. Each of the two types of propagating species preferentially adds the other monomer; that is, M1*,M2* adds only M2,M1, and an alternating arrangement of monomers is incorporated along the copolymer. Most comonomer systems lie between ideal and alternating behaviors, with an increasing tendency toward alternation as r_1 and r_2 decrease from unity toward zero. This is illustrated in Figure 3.6. For r_1 and r_2 both <1 the F_1 vs. f_1 curves cross the $f_1 = F_1$ line. At these crossover points the feed and copolymer compositions are identical, so $f_1 = (1 - r_1)/(2 - r_1 - r_2)$ results from the copolymer equation and copolymerization occurs without a change in feed composition. These copolymerizations are termed *azeotropic*.

Finally if both r_1 and r_2 exceed unity, then there is a tendency to form copolymers with a blocky incorporation of monomers. This behavior is quite rare; as a consequence, block copolymers are prepared in another manner called *living chain-growth polymerization*, which employs ionic initiators and will be discussed subsequently.

EMULSION POLYMERIZATION

We have seen that in a typical chain-growth polymerization initiated by free radicals the rate of polymerization R_p and the size of the polymers formed v are directly and inversely proportional, respectively, to $[I]^{1/2}$. Thus we are unable to speed up the production without decreasing the molecular weight of the polymers formed by increasing the initiator concentration in a homogeneous chain-growth polymerization. However, their exists a means for conducting free radical chain-growth polymerizations where both R_p and v can be simultaneously increased or decreased, and this process is called *emulsion polymerization* (Harkins, 1947; Smith and Ewart, 1948; Odian, 1991).

When a typical vinyl monomer $CH_2=\underset{\underset{X}{|}}{CH}$ is added to water along with a surfactant or emulsifier, an emulsion of monomer stabilized by the surfactant is produced (see Figure 3.7). If a water-soluble free radical initiator is added to this emulsion, it becomes possible to conduct an emulsion chain-growth polymerization. Above a surfactant concentration called the *critical micelle concentration* (CMC), the

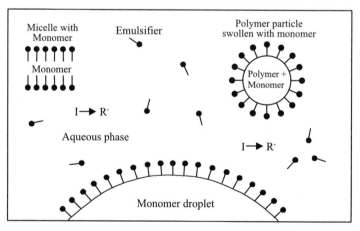

Figure 3.7. Simplified representation of an emulsion polymerization system.

excess surfactant aggregates into bilayer micelles (see Figure 3.7), which together with the monomer droplets contain nearly all of the hydrophobic monomers. When a free radical moves into a monomer micelle, chain-growth polymerization is initiated, and the micelle is soon converted into a polymer particle swollen with monomer (see Figure 3.7). As monomer micelles are converted to polymer particles, new monomer micelles are created at the expense of the monomer droplets, because inititiation and growth of polymer occurs preferentially in the monomer micelles since their surface to volume ratio far exceeds that of the monomer droplets.

It is easy to understand that the overall rate of polymerization would be expected to increase for a given initiator concentration if we simply increased the number of monomer micelles, which after initiation soon become polymer particles. Increasing the surfactant concentration will increase the monomer micelle concentration, leading to larger R_p. The number-average degree of polymerization $\bar{X}_n$ would be expected to be given by the ratio of the rate of growth of the polymer in a polymer particle to the rate of entry of primary radicals into the polymer particle, that is, $\bar{X}_n = r_p/r_i$. $r_p = k_p[M]$ and $r_i = R_i/N$, where N is the number of polymer particles and R_i is the rate of free radical initiation. Thus, $\bar{X}_n = (Nk_p[M])/R_i$, or $\bar{X}_n$ is directly proportional to the number of polymer particles, which may be increased by adding more surfactant. Clearly both R_p and $\bar{X}_n$ can be made to simultaneously increase or decrease by adjusting the surfactant concentration employed in an emulsion polymerization (Smith and Ewart, 1948).

LIVING CHAIN-GROWTH POLYMERIZATION

Nearly all monomers containing carbon–carbon double bonds undergo free-radical-initiated chain-growth polymerization. However, because of strict requirements for stabilization of anionic and cationic propagating species, ionic polymerizations are highly selective. This high monomer selectivity makes the commercial utilization of cationic and anionic polymerizations rather limited. Ionic polymerizations are not as well understood as free radical polymerizations, because of experimental difficulties such as the use of heterogeneous inorganic initiators, very rapid polymerization rates, and extreme sensitivity to

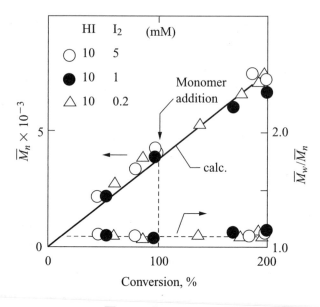

Figure 3.8. Dependence of $\overline{M}_n$ and $\overline{M}_w/\overline{M}_n$ on conversion for the polymerization of i-butyl vinyl ether by HI/I$_2$ in CH$_2$Cl$_2$ at $-15°$C. [M] $= 0.38$ M at beginning of each batch; [HI] $= 0.01$ M; [I$_2$] $= 0.02$ M ($\triangle$), 0.001 M ($\bullet$), 0.005 M ($\circ$). After Sawamoto and Higashimura (1986) (by permission of Huthig and Wepf Verlag, Basel).

impurities. For these reasons we will limit our discussion of ionic chain-growth polymerization to living polymerizations (Swarz, 1968), which have been widely used to make block copolymers.

In living polymerizations the propagating centers do not undergo either termination or transfer. There are several ways to determine whether or not a polymerization is living. In a highly purified reaction system spectroscopic observation of the propagating species persists throughout the polymerization without disappearance or decrease in intensity even at 100% conversion. A plot of $\overline{M}_n$ or $\overline{X}_n$ versus conversion is linear, because the number of propagating species does not change throughout the polymerization. After 100% conversion, additional polymerization takes place following the addition of more monomer. These later characteristics are illustrated by the data presented in Figure 3.8, where the cationic polymerization of isobutyl vinyl ether clearly conforms to a living polymerization. Note that the PDI $= \overline{M}_w/\overline{M}_n$ remains constant at ~ 1.0 throughout the polymerization, because all growing, living chains are initiated almost simultaneously at the beginning of the polymerization. If after 100% conversion a monomer other than isobutyl vinyl ether, which also undergoes cationic chain-growth polymerization, were added instead, then clearly a block copolymer would result. This is in fact a method for producing block copolymers (Noshay and McGrath, 1977).

Living ionic polymerizations are important for their ability to produce various block copolymers and because they yield several homopolymers with controllable molecular weights and very narrow molecular weight distributions (PDI ~ 1). In Chapter 7, where we discuss the bulk properties of polymers, it will be observed that both the molecular weight and molecular weight distribution can have important effects on bulk polymer properties.

DISCUSSION QUESTIONS

1. Compare and contrast typical step- and chain-growth polymerizations in terms of their monomer reactivities, reaction conditions, and the contents and molecular weights of polymers formed as the % monomer conversion increases.

2. Given the vinyl monomer $CH_2{=}CH$ and the free radical initiator
$$\underset{Cl}{|}$$

$$\overset{\displaystyle CH_3 \qquad CH_3}{\underset{\displaystyle CH_3 \qquad CH_3}{CH_3{-}C{-}O{-}O{-}C{-}CH_3,}}$$ illustrate the chain-growth polymerization of poly(vinyl chloride). Be sure to include initiation, propagation, and termination steps.

3. Increasing the concentration of initiator used in a chain-growth polymerization has two main consequences. What are they and why do they occur?

4. A free radical chain-growth polymerization produces polymers that contain an average of 1.2 initiator fragments. Assuming no chain-transfer has occurred during the polymerization, what are the relative extents of termination by coupling and by disproportionation?

*5. When a chain-growth polymer is terminated by disproportion-
ation, why do we call one of the terminated chains "dead" and
the other "dormant"?

*6. Why is it not possible to predict the comonomer sequence distri-
bution for an M1,M2 copolymer from the rates of polymerization
observed in the homopolymerizations of poly-M1 and poly-M2?

*7. When copolymerized, acrylonitrile (1) and acrylamide (2) have
$r_1 = 0.86$ and $r_2 = 0.81$. Discuss their comonomer sequences for a
50:50 comonomer feed ratio. Do the same for the acrylonitrile/
isobutylene pair, where $r_1 = 0.98$ and $r_2 = 0.020$.

*8. Why are both the rates and the degrees of polymerization or the
molecular weights increased in an emulsion polymerization when
the surfactant concentration is increased?

*9. When the molecular weight of the chain-grown polymer described
in Question 4 is measured, it is found to be considerably less than
expected from the initial amount of monomer employed and the
total number of initiator fragments terminating the chains in the
synthesized polymer sample. What is likely occurring during this
chain-growth polymerization? When this same polymerization
is conducted in solution, the molecular weight of the resulting
polymer is found to increase compared with a sample polymerized in bulk. Does this observation suggest any further, more
detailed features of the polymerization, and, if so, why and what
are they?

REFERENCES

Alfrey, T., and Goldfinger, G. (1944), *J. Chem. Phys.*, **12**, 115, 205, 332.

Flory, P. J. (1953), *Principles of Polymer Chemistry*, Cornell University
Press, Ithaca, New York, Chapters IV and VIII.

Harkins, W. D. (1947), *J. Am. Chem. Soc.*, **69**, 1428.

Hart, H., Hart, D. J., and Crain, L. E. (1995), *Organic Chemistry: A Short
Course*, Houghton Mifflin, Boston, Chapter 3.

Kondratiev, V. N. (1969), Chain Reactions, Chapter 2 in *Comprehensive
Chemical Kinetics*, Vol. 2, Bamford, C. H., and Tipper, C. F. H., Eds.,
American Elsevier, New York.

Landler, Y. (1950), *Compt. Rend.*, **230**, 532.

Mayo, F. R., and Lewis, F. M. (1944), *J. Am. Chem. Soc.*, **66**, 1594.

Mayo, F. R., and Walling, C. (1950), *Chem. Rev.*, **46**, 191.

Morgan, P. W., and Kwolek, S. L. (1959), *J. Chem. Educ.*, **36**, 182 and cover; *J. Polym. Sci.*, **40**, 299.

North, A. M. (1974), The Influence of Chain Structure on the Free Radical Termination Reaction, Chapter 5 in *Reactivity, Mechanism, and Structure in Polymer Chemistry*, Jenkins, A. D., and Leswith, A. Eds., John Wiley and Sons, New York.

Noshay, A., and McGrath, J. E. (1977), *Block Copolymers*, Academic Press, New York.

Odian, G. (1991), Principles of Polymerization, 3rd ed., John Wiley and Sons, New York, Chapters 3–6.

Pepper, D. C. (1954), *Quart. Rev.* (London), **8**, 88.

Rodriguez, F., Mathias, L. J., Kroschwitz, J., and Carraher, C. E., Jr. (1987), *J. Chem Educ.*, **64**, 886.

Santee, G. F., Marchessault, R. H., Clark, H. G., Kearny, J. J., and Sawamoto and Higashimura (1986), *Makromol. Chem. macromol. Sympos.* **3**, 83.

Stannett, V. (1964), *Makromol. Chem.*, **73**, 177.

Schulz, G. V., and Blaschke, F. (1942), *Z. Physik (Leipzig)*, **B51**, 75.

Smith and Ewart (1948), *J. Chem. Phys.* **16**, 592.

Sugimura, T., and Minoura, Y. (1966), *J. Polym. Sci.*, **A-1**(4), 2735.

Swarz, M. (1968), *Carbanions, Living Polymers, and Electron-Transfer Processes*, Wiley-Interscience, New York.

Tedder, J. M. (1974), The Reactivity of Free Radicals, Chapter 2 in *Reactivity, Mechanism, and Structure in Polymer Chemistry*, Jenkins, A. D., and Ledwith, A., Eds., Wiley-Interscience, New York.

Vrancher, A. and Smets, G. (1959), *Makromol. Chem.*, **30**, 197.

Wall, F. T. (1944), *J. Am. Chem. Soc.*, **66**, 2050.

Walling, C. (1957), *Free Radicals in Solution*, John Wiley and Sons, New York.

CHAPTER 4

THE MICROSTRUCTURES OF POLYMERS

INTRODUCTION

The microstructure of a polymer is the connectivity and bonding of atoms and groups of atoms forming its backbone and the side chains attached to it. Polymer microstructure is usually determined during the course of polymerization as the monomers are inserted into the growing polymer chain. Because of the predictable nature of step-growth polymerizations, where monomers can only be linked through reaction of their functional groups, we generally need consider only the molecular weights and not the microstructures of the resultant polymers. Rather, the directions of monomer addition, the stereochemical and geometric isomeric forms of the incorporated monomers, and the order of addition of comonomer units are the possible elements of microstructure which may be introduced during a chain-growth polymerization. We will briefly illustrate these microstructural features using vinyl and diene polymers as examples.

On a larger scale, sometimes as a direct result of polymerization and at other times due to postpolymerization chemical reactions, the linear architecture of polymers can be modified to yield branched and cross-linked structures. The types, lengths, and locations of the branches and cross-links can also be considered components of a polymer's microstructure.

We will present a brief discussion of the use of NMR spectroscopy to experimentally determine polymer microstructure, and we will attempt to relate the physical properties exhibited by polymers to their microstructures. Here we stress the wide range of properties observed for polymers and their underlying causes, the rich variety of possible microstructures.

ELEMENTS OF POLYMER MICROSTRUCTURE

Let us consider the types of polymer microstructures, which can be produced by the isomerism of monomers during their incorporation into the growing polymer chain:

$$n \overset{A}{\underset{B}{\diagdown}} C = C \overset{H}{\underset{H}{\diagup}} \longrightarrow -(-\overset{A}{\underset{B}{\overset{|}{C}}} - \overset{H}{\underset{H}{\overset{|}{C}}} -)_n -$$

If both A and B are not H in the vinyl polymerization indicated above, then each monomer unit can be potentially enchained in either

of two directions, that is, $-\overset{A}{\underset{B}{\overset{|}{C}}}-CH_2-$ or $-CH_2-\overset{A}{\underset{B}{\overset{|}{C}}}-$, leading to among

others the following regiosequences:

(1)
$$-\overset{A}{\underset{B}{\overset{|}{C}}}-CH_2-\overset{A}{\underset{B}{\overset{|}{C}}}-CH_2-\overset{A}{\underset{B}{\overset{|}{C}}}-CH_2-\overset{A}{\underset{B}{\overset{|}{C}}}-CH_2-\overset{A}{\underset{B}{\overset{|}{C}}}-CH_2-$$

(2)
$$-\overset{A}{\underset{B}{\overset{|}{C}}}-CH_2-\overset{A}{\underset{B}{\overset{|}{C}}}-CH_2-CH_2-\overset{A}{\underset{B}{\overset{|}{C}}}-\overset{A}{\underset{B}{\overset{|}{C}}}-CH_2-\overset{A}{\underset{B}{\overset{|}{C}}}-CH_2-$$

Note that in (1) all monomer units have been added in the same direction, producing a regioregular structure, while in (2) the third monomer has been inserted in the inverted direction, resulting in a regioirregular structure. Let us assume that all monomers

$$\begin{matrix} A & & H \\ & \diagdown & \diagup \\ & C=C \\ & \diagup & \diagdown \\ B & & H \end{matrix}$$

are added in the same direction during their polymerization (regioregular addition). There still remains a degree of structural freedom that becomes fixed during the polymerization, that is, the stereochemical configuration or relative handedness of successive monomer units. Disregarding chain ends, the main-chain-substituted carbons do not have the necessary four different substituents to qualify as asymmetric centers, but they do have the opportunity for relative handedness and are therefore termed *pseudoasymmetric*. The three structures below illustrate various possible stereochemical arrangements produced by the polymerization of monomers possessing a pseudoasymmetric carbon:

1. *Isotactic*

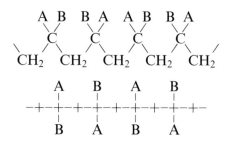

2. *Syndiotactic*

3. *Atactic*

$$
\begin{array}{cccc}
\text{A B} & \text{B A} & \text{B A} & \text{A B} \\
\diagdown\diagup & \diagdown\diagup & \diagdown\diagup & \diagdown\diagup \\
\text{C} & \text{C} & \text{C} & \text{C} \\
\diagup\diagdown & \diagup\diagdown & \diagup\diagdown & \diagup\diagdown \\
\text{CH}_2 & \text{CH}_2 \;\; \text{CH}_2 & \text{CH}_2 & \text{CH}_2
\end{array}
$$

$$
\begin{array}{cccc}
\text{A} & \text{B} & \text{B} & \text{A} \\
| & | & | & | \\
-+\!-\!+\!-\!+\!-\!+\!-\!+\!-\!+\!-\!+\!-\!+\!-\!+\!- \\
| & | & | & | \\
\text{B} & \text{A} & \text{A} & \text{B}
\end{array}
$$

   Each of the polymer chain fragments has been drawn in the planar, zigzag, or all-*trans* conformation to provide a clear perspective of the handedness of stereochemical arrangements of neighboring repeat units. The two stereoregular structures (1) and (2), where the substituents A and B are all either on the same side of the zigzag backbone plane or alternate from side to side, are called *isotactic* and *syndiotactic*, respectively. The *atactic* structure (3) is characterized by an irregular, random arrangement of neighboring substituent groups on either side of the backbone.
   Polymerization of diene monomers, such as butadiene (CH_2=CH–CH=CH_2), can produce polymers with varying geometrical structures. 1,4-Enchainment of monomers can produce *cis*(Z) and *trans*(E) geometrical isomers, which remain stable in the polymer because their interconversion by rotation about the double bond is resisted and generally prevented by the necessity of breaking the π-bond during rotation. 1,2-Enchainment leads to structures with the same configurational properties as vinyl polymers:

$$
\begin{array}{cc}
\begin{array}{c}
\text{H} \qquad \text{H} \\
\diagdown \qquad \diagup \\
\text{CH=CH} \\
\diagup \qquad \diagdown \\
-\text{CH}_2 \qquad \text{CH}_2-
\end{array}
&
\begin{array}{c}
\text{H} \qquad \text{CH}_2- \\
\diagdown \qquad \diagup \\
\text{CH=CH} \\
\diagup \qquad \diagdown \\
-\text{CH}_2 \qquad \text{H}
\end{array} \\
cis(Z) & trans(E)
\end{array}
$$

$$
\begin{array}{c}
\text{CH}_2 \\
\|\\
\text{CH} \\
|\\
-\text{CH}_2-\text{CH}-\text{CH}_2-\text{CH}-\text{CH}_2-\text{CH}-\text{CH}_2-\text{CH}-\text{CH}_2- \\
\qquad\quad |\qquad\qquad\quad |\qquad\qquad\qquad\qquad\quad | \\
\qquad\quad \text{CH}\qquad\qquad \text{CH}\qquad\qquad\qquad\qquad \text{CH} \\
\qquad\quad \|\qquad\qquad\quad \|\qquad\qquad\qquad\qquad\quad \| \\
\qquad\quad \text{CH}_2\qquad\qquad \text{CH}_2\qquad\qquad\qquad\qquad \text{CH}_2
\end{array}
$$

(Atactic 1,2-polybutadiene)

Natural rubber is an example of a polymer obtained through polymerization of a 2-substituted butadiene, $CH_2=\overset{\overset{\displaystyle X}{|}}{C}-CH=CH_2$, where $X = CH_3$ and all units are enchained in the 1,4-*cis*(Z) arrangement. The 1,4-*trans*(E) enchainment results in balata or gutta percha, which is produced from a different plant. 2-Substituted butadienes may also be incorporated into the polymer chain in two different directions, as indicated below. Thus, regardless of the mode of enchainment (1,4-*cis* or -*trans*, 1,2- or 3,4-) different regiosequences may also be produced during their polymerization.

Head-to-Tail

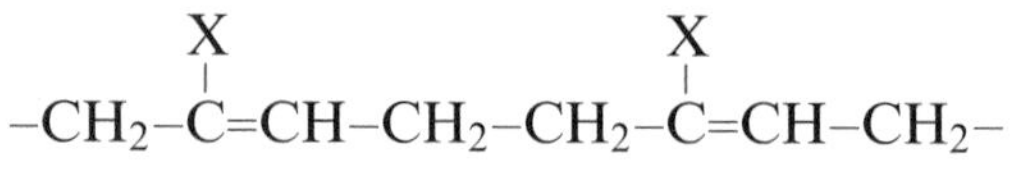

$$-CH_2-\overset{\overset{\displaystyle X}{|}}{C}=CH-CH_2-CH_2-\overset{\overset{\displaystyle X}{|}}{C}=CH-CH_2-$$

Head-to-Head, Tail-to-Tail

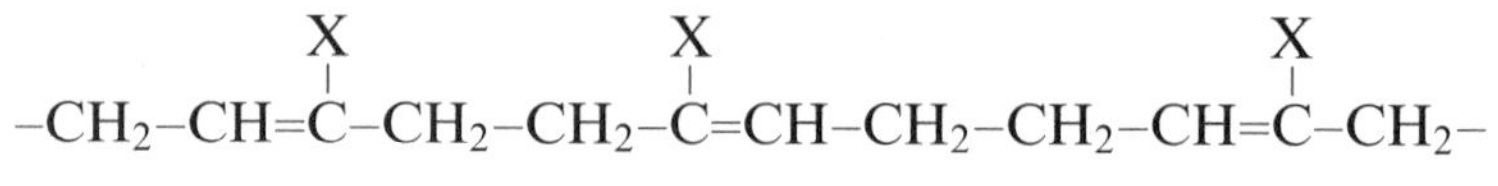

$$-CH_2-CH=\overset{\overset{\displaystyle X}{|}}{C}-CH_2-CH_2-\overset{\overset{\displaystyle X}{|}}{C}=CH-CH_2-CH_2-CH=\overset{\overset{\displaystyle X}{|}}{C}-CH_2-$$

Of course, similar regiosequences are possible for 1,2- and 3,4- modes of monomer enchantment:

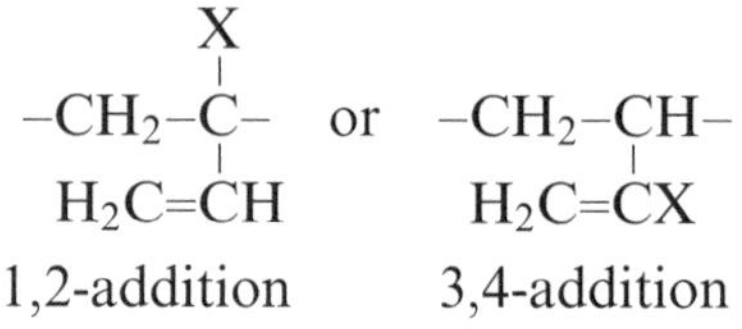

$$-CH_2-\overset{\overset{\displaystyle X}{|}}{\underset{\underset{\displaystyle H_2C=CH}{|}}{C}}- \quad or \quad -CH_2-\overset{}{\underset{\underset{\displaystyle H_2C=CX}{|}}{CH}}-$$

1,2-addition   3,4-addition

Even the deceptively simple structure of a 2-substituted butadiene monomer can lead to a polymer whose microstructure is very complex. At the three monomer unit or triad level, nearly 300 different microstructures are possible when each monomer unit can be enchained by 1,4-*cis*(Z) or *trans*(E), 1,2-addition, or 3,4-addition; in either of two directions (head-to-tail or tail-head) and for those triads containing at least two monomer units enchained by 1,2 and/or 3,4-addition, different stereosequences (m or r; and mm, mr, rm, and rr), where m,r denote meso, racemic diads with the same, opposite attachment of side chains. It is left to the reader to confirm the validity of this statement concerning the microstructures that are possible for poly-2-substituted butadienes. Vinyl monomers with true asymmetric centers in their side chains produce polymers with asymmetric side chains. If one of the optically active monomeric enantiomers (D or L, or R or S) is polymerized, the resulting polymer will be optically active and may also be isotactic, syndiotactic, or atactic in the usual sense. Though vinyl monomers cannot be polymerized into polymers with truly asymmetric main-chain carbons, other monomers, such as propylene oxide, do produce such polymers (Pruitt and Baggett, 1955; Price and Osgan, 1956; Osgan and Price, 1959; Tsurata, 1967):

$$
n \quad \begin{array}{c} CH_3 \quad H \\ \diagdown \diagup \\ C\text{-}C \quad (R, S) \\ \diagup \diagdown \diagup \diagdown \\ H \quad O \quad H \end{array}
$$

$$
\begin{array}{ccccc}
CH_3 & & H & & CH_3 \\
| & & | & & | \\
-C\text{-}CH_2\text{-}O\text{-}C\text{-}CH_2\text{-}O\text{-}C\text{-}CH_2\text{-}O- \\
| & & | & & | \\
H & & CH_3 & & H
\end{array}
$$

(Isotactic, *RRR* or *SSS*)

$$
\begin{array}{ccccc}
CH_3 & & CH_3 & & CH_3 \\
| & & | & & | \\
-C\text{-}CH_2\text{-}O\text{-}C\text{-}CH_2\text{-}O\text{-}C\text{-}CH_2\text{-}O- \\
| & & | & & | \\
H & & H & & H
\end{array}
$$

(Syndiotactic, *RSR* or *SRS*)

[Note that unlike vinyl polymers whose pseudoasymmetric carbons are separated by two bonds, the truly asymmetric carbons in poly-(propylene oxide) are separated by three bonds and so the methyl carbons in meso diads or the isotactic chain alternate on opposite sides of the backbone, while they are attached to the same side of the backbone in racemic diads in the syndiotactic chain.]

Because the poly(propylene oxide) chains have a sense of direction, there are two heterotactic structures that cannot be superimposed (non-mirror images):

$$\begin{array}{ccc}
CH_3(H) & H(CH_3) & H(CH_3) \\
| & | & | \\
-C-CH_2-O-C-CH_2-O-C-CH_2-O- \\
| & | & | \\
H(CH_3) & CH_3(H) & CH_3(H)
\end{array}$$

[Heterotactic-1(2), *RRS*(*SRR*) or *SSR*(*RSS*)]

In nature the most prevalent examples of truly asymmetric polymers are the polypeptides and proteins $-(-NH-\overset{\overset{\textstyle R}{|}}{CH}-\overset{\overset{\textstyle O}{\|}}{C}-)_n-$ (see Chapter 8). In proteins the peptide residues are invariably of the L-configuration, while in small, usually cyclic polypeptides which function as hormones, toxins, antibiotics, or ionophores, both L and D residues are found (Tonelli, 1986).

COPOLYMER SEQUENCES

Though to this point we have limited our discussion of polymer microstructure to homopolymers obtained from the polymerization of a single monomer, as is common in nature (e.g., proteins and polynucleic acids), two or more different monomer units may be incorporated, resulting in a copolymer. The comonomer units may be enchained randomly, in regular alternation, or in block or graft structures:

Random−o−o−●−o−●−o−●−●−●−o−o−●−o−●−●−o−●−o−

Alternating−o−●−o−●−o−●−o−●−o−●−o−●−o−●−o−●−o−●−o−

Block−o−o−o−o−o−o−o−o−o−o−●−●−●−●−●−●−●−●−●−●−

Graft−o−o−o−o−o−o−o−o−o−o−o−o−o−o−o−o−o−o−o−o

Aside from the comonomer sequence and method of attachment, co-polymer microstructures are also subject to the effects of stereo- and regiosequence as discussed previously. Clearly the microstructures of homo- and copolymers are varied and afford the synthetic chemist an almost limitless number of variations on a macromolecular theme.

ORGANIZATION OF POLYMER CHAINS

Though we may have implied otherwise, polymer chains are not always linear. Polymers may contain branches produced during their polymerization either through chain transfer of radicals or by utilizing trifunctional monomers, and they may be grafted onto their backbones by postpolymerization reaction. Cross-linked polymer samples may be produced by polymerization with multifunctional monomers (trifunctional and higher) or by postpolymerization reaction or by irradiation. At sufficiently high degrees of cross-linking, a three-dimensional network can be formed.

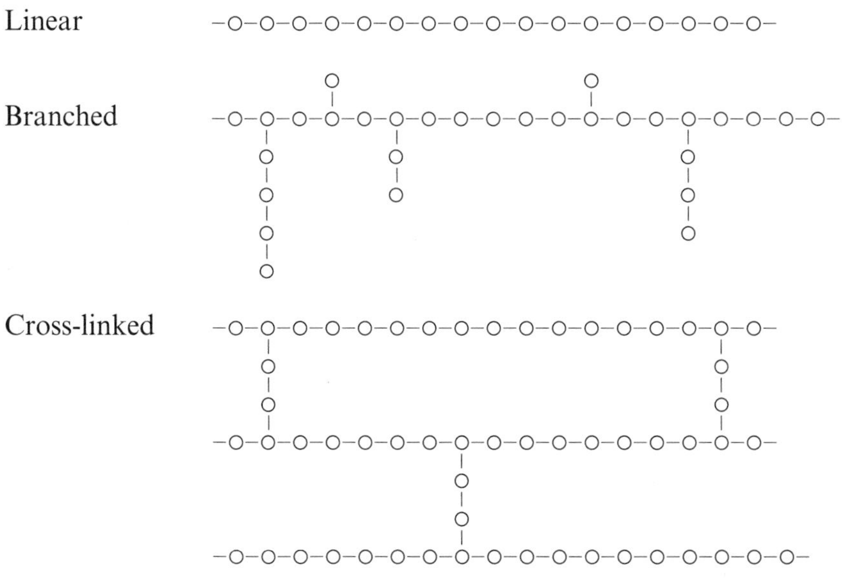

Linear

$-$O$-$O$-$O$-$O$-$O$-$O$-$O$-$O$-$O$-$O$-$O$-$O$-$O$-$O$-$O$-$O$-$O$-$

Branched

Cross-linked

POLYMER MICROSTRUCTURES AND PROPERTIES

Polymer scientists are motivated to study the microstructures of polymers by a desire to understand their amazingly varied and often unique physical properties. Why are some polymers rigid and strong, while others are elastic and deformable and may even flow at the same temperature? Why do some polymers resist degradation caused by exposure to heat, chemicals, and radiation, while others degrade rapidly? Why do some polymers become brittle and shatter on impact at low temperatures, while others remain tough and resistant to impact under the same conditions? What is it about the microstructures of proteins (the primary sequence of amino acid or peptide residues) that make some proteins function as the structural components of teeth, bones, skin, and hair, while other proteins function as enzymes to catalyze the body's biochemical reactions? Answers to these and similar questions are most likely to be found in a detailed understanding of polymer microstructure. Though a main theme that we will further develop throughout this book, let us briefly introduce here some connections between polymer properties and polymer microstructures. It is well known (Ward, 1985) that the strongest polymers are those whose linear chains can be highly oriented into fibers where they crystallize. The ability of a polymer to crystallize depends on its microstructure. For example, polypropylene $[-(-CH_2-\overset{\overset{\displaystyle CH_3}{|}}{CH}-)_n-](PP)$, when polymerized with an appropriate catalyst, is highly isotactic (i). The stereoregular enchainment of pendant methyl groups in i-PP permits the facile packing and crystallization of its chains ($T_{\text{melt}} \sim 200°C$). When crystalline i-PP chains are drawn, strong fibers are produced, which can be used (for example) to produce strong, lightweight rope. Atactic PP chains, with their stereoirregular distribution of methyl substituents, cannot pack tightly and crystallize, so a-PP is an amorphous polymer that slowly creeps under stress.

When propylene and ethylene are copolymerized, amorphous E-P copolymers are produced at intermediate comonomer compositions. When subsequently cross-linked, they result in a commercially important class of synthetic rubbers. E-P copolymers rich in ethylene are found to crystallize like PE homopolymer, but to a lesser degree and with a greatly reduced content of long-chain branches. These copolymers are called linear, low-density PE and are much easier to melt-process than PE homopolymers.

Amorphous polymers exhibit widely different bulk properties that are very temperature-dependent. Below 100°C, atactic polystyrene $[-(CH_2-CH-)_n-](PS)$ is rigid and, as we all know, serves well as a

material to contain hot beverages in the form of styrofoam cups. At higher temperatures, PS begins to lose its shape and flow. The transition temperature between rigid (glassy) and soft (rubbery) behavior is termed the glass-to-rubber transition temperature, T_g (Ferry, 1970); $T_g = 100°C$ for PS.

In Figure 4.1 the glass transition temperatures of a series of vinyl-

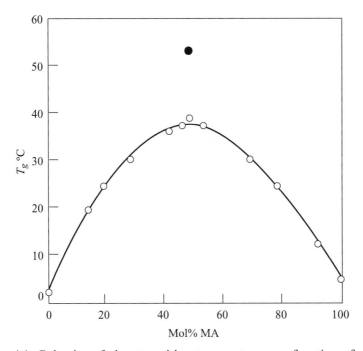

Figure 4.1. Behavior of glass transition temperature as a function of comonomer composition for VDC–MA copolymers (Hirooka and Kato, 1974): ○, random; ●, regularly alternating. [Reprinted with permission from Tonelli (1975).]

idene chloride $[-(-CH_2-\overset{\overset{\text{Cl}}{|}}{\underset{\underset{\text{Cl}}{|}}{C}}-)_n-]$-methyl acrylate-$[(-CH_2-\overset{}{\underset{\underset{O=C-O-CH_3}{|}}{CH}}-)_n-]$

(VDC–MA) copolymers are presented as a function of their overall comonomer composition. Clearly the T_g's of VDC–MA copolymers are elevated above those of their constituent homopolymers. Even more interesting is the observation that regularly alternating VDC–MA has a T_g considerably higher than a 50:50 VDC–MA copolymer with a random distribution of comonomer units. Both copolymers have the same overall comonomer composition but differ in their microstructures—that is, their comonomer sequence distributions. [See Tonelli (1975) and Chapter 7 for a possible explanation of this behavior.]

Having illustrated only several among the myriad possible polymer microstructures and having suggested how in a few instances they might influence the physical properties of polymers, in the remaining portion of this chapter we present a very brief account of how nuclear magnetic resonance spectroscopy can be used to unravel their microstructural details.

DETERMINATION OF POLYMER MICROSTRUCTURE-NMR SPECTROSCOPY

Because the detailed connectivity of atoms along and attached to the backbones of polymers can sensitively affect their physical properties, as we shall soon demonstrate, experimental determination of polymer microstructures is the first and most important order of business in the process of their characterization. So how are we to probe the microstructures of high-molecular-weight polymer samples whose constituent repeat units may be incorporated with different regio- and stereosequences, different geometrical arrangements, and different comonomer sequences and may by design or chain transfer result in branched and cross-linked structures as well? The answer is nuclear magnetic resonance (NMR) spectroscopy. The reason that the NMR technique can be successfully employed to unravel the microstructures of polymers is because the underlying resonance phenomenon is exquisitely sensitive to the detailed structural architecture of molecules including polymers.

Though the nuclei of all atoms possess charge (protons) and mass (protons and neutrons), only those nuclei with odd mass or charge numbers also have angular momentum and a magnetic moment. The former nuclei have spin angular-momentum quantum numbers I whose values are odd integral multiples of $1/2$, while the nuclei with even mass and odd charge numbers have integral spin 1. The angular momentum of a nucleus with spin I is simply I $(\hbar/2\pi)$, where $\hbar$ is Planck's constant. These nuclei possess a magnetic moment μ, which is taken parallel to the angular momentum vector. The magnetic moment vector takes on only certain values along any chosen axis which are defined by a series of magnetic quantum numbers $m = I, I - 1, I - 2, \ldots, -I$. For nuclei of interest here—that is, those found most often in polymers (^{1}H, ^{13}C, ^{15}N, ^{19}F, ^{29}Si, ^{31}P)—we have $I = 1/2$, and thus $m = +1/2$ and $-1/2$. There are $2I + 1$ possible orientations of μ, or magnetic states. The ratio of their magnetic moment to their angular momentum is called the magnetogyric ratio $\gamma = 2\pi\mu/\hbar I$ and is characteristic of each nucleus.

Both nuclear magnetic states have the same energy in the absence of a magnetic field and are thus degenerate, or equally populated, but upon application of a uniform magnetic field B_0 they correspond to states of different potential energy. The magnetic moment μ is aligned either along ($m = +1/2$) or against ($m = -1/2$) the field B_0, with the latter state corresponding to a higher energy. Detection of the transitions of the magnetic nuclei between these spin states [$m = +1/2$ (parallel), $m = -1/2$ (antiparallel)] are made possible by the NMR phenomenon.

In Figure 4.2 we have drawn a nuclear magnetic moment μ in the presence of a static applied magnetic field B_0 acting along the z-axis of the coordinate system. The angle θ between the magnetic moment and the applied magnetic field does not change, because the torque, $L = \mu \times B_0$, tending to tip μ toward B_0 is exactly balanced by the spinning of the magnetic moment, resulting in nuclear precession about the z-axis. Attempting to force the alignment of μ along the z-axis by increasing B_0 only results in faster precession. A good macroscopic analogy is afforded by observation of a spinning top in the earth's gravitational field.

The precessional or Larmor frequency, ν_0, of the spinning nucleus is given by $\gamma B_0/2\pi$ and is independent of θ. However, the energy of the spin system does depend on the angle between μ and B_0: $E =$

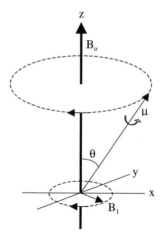

Figure 4.2. Nuclear magnetic moment μ in a magnetic field B_0.

$-\mu B_0 = -\mu B_0 \cos \theta$. We may change the orientation θ between μ and B_0 by application of a small rotating magnetic field B_1 orthogonal to B_0 (see Figure 4.2). Now μ will experience the combined effects of both B_1 and B_0 if the angular frequency of B_1 coincides with v, the precessional frequency of the spin. In this situation the nucleus absorbs energy from B_1 and θ changes: otherwise B_1 and μ would not remain in phase and no energy would be transferred between them.

If the rotation rate of B_1 is varied through the Larmor frequency of the nucleus, a resonance condition is achieved, accompanied by a transfer of energy from B_1 to the spinning nucleus and an oscillation of the angle θ between B_0 and μ. At $B_0 = 2.34$ T (1 T $= 1$ tesla $= 10$ kilogauss) the resonant or Larmor fequencies of the ^{1}H, ^{19}F, ^{31}P, ^{13}C, ^{29}Si, and ^{15}N nuclei are $v_0 = 100, 94, 40.5, 25.1, 19.9,$ and 10. 1 MHz, respectively. However, if all nuclei of the same type (e.g. all protons) were to resonate and be flipped by rotating B_1 at the same Larmor frequency v_0, then NMR would not be a spectroscopic tool useful for the study of molecular structure, but would be limited instead to providing compositional information, such as the ratios C:H:N:F, and so on, without illucidating patterns of atom connectivity. Fortunately, it is observed that the characteristic resonant or Larmor frequency of a nucleus depends sensitively upon its chemical and/or structural environment (Tonelli, 1989).

The cloud of moving electrons about each nucleus produces orbital currents when placed in a magnetic field B_0. These currents produce small local magnetic fields, which are proportional to B_0 but are opposite in direction, thereby effectively shielding the nucleus from B_0. Consequently, a slightly higher value of B_0 is needed to achieve resonance. The actual local field B_{loc} experienced by a nucleus can be expressed as $B_{loc} = B_0(1 - \sigma)$, where σ is the screening constant. σ is highly sensitive to chemical structure but independent of B_0. The resonant Larmor frequency becomes $\nu_0 = \mu B_{loc}/\hbar I = \mu B_0(1 - \sigma)/\hbar I$. Nuclear shielding is influenced by the numbers and types of atoms and groups attached to or near the observed nucleus. The dependence of σ upon molecular structure lies at the heart of NMR's utility as a probe of molecular structure including the microstructures of polymers.

There is no fundamental scale unit in NMR, nor is there a natural zero of reference. These difficulties are avoided by expressing the resonant frequencies of nuclei in parts-per-million (ppm) relative changes in B_0 and referring these changes or displacements in reso- nance, called chemical shifts (δ), to the ppm relative change in the resonant frequency of an arbitrary reference substance added to the sample. In ^{1}H and ^{13}C NMR spectroscopy, for example, it is cus- tomary to use tetramethylsilane (TMS) as the reference compound, where the chemical shifts of both the ^{1}H and ^{13}C nuclei of TMS are taken as $\delta = 0.0$ ppm.

$$\text{TMS} = \text{CH}_3\text{-}\underset{\underset{\text{CH}_3}{|}}{\overset{\overset{\text{CH}_3}{|}}{\text{Si}}}\text{-CH}_3$$

We have seen that the magnetic field B_i required to obtain the resonance condition for nucleus i at a particular irradiating field frequency B_1 is not equal to the applied field B_0, but is instead $B_i = B_{loc} = B_0(1 - \sigma)$, where the nuclear screening constant σ depends on the chemical structural environment of nucleus i. By producing small local magnetic fields, the cloud of electrons moving about the nucleus shield the nucleus from the applied field B_0. Any structural feature that alters the electronic environment of a nucleus will affect its screening constant σ and lead to an alteration in its chemical shift δ at resonance.

CONNECTING POLYMER MICROSTRUCTURES WITH THEIR NMR SPECTRA

An illustration of the microstructural sensitivity of the NMR spectrum of a polymer is presented in Figure 4.3, where the ^{13}C NMR spectra for isotactic, atactic, and syndiotactic polypropylene (PP) are

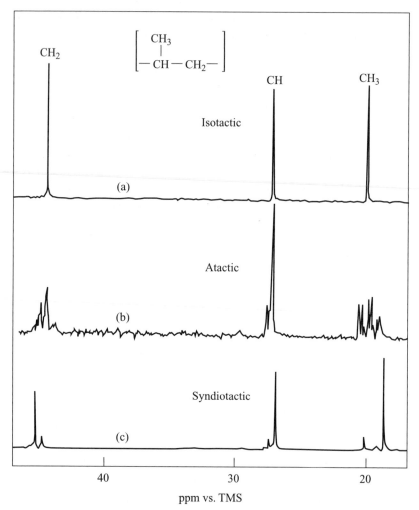

Figure 4.3. 25-MHz ^{13}C NMR spectra of (a) isotactic, (b) atactic, and (c) syndiotactic PP. [Reprinted with permission from Tonelli and Schilling (1981).]

displayed (Tonelli and Schilling, 1981). Note the clear separation between the CH_2, CH, and CH_3 resonant frequencies exhibited by each PP sample. Careful observation also reveals that the resonant frequencies of all three-carbon types are not coincident, however, among the samples. For example, the chemical shifts of the CH_2 and CH_3 carbons in s-PP (syndiotactic) are displaced down- and upfield, respectively, from their resonant frequencies in the isotactic sample (i-PP). Furthermore, in the atactic sample (a-PP) all three-carbon types evidence multiple resonances. The microstructural differences between i-, a-, and s-PP are clearly evident in their different ^{13}C NMR spectra.

What are the microstructural differences between i-PP, a-PP, and s-PP? Each possesses a different stereosequence, with all meso or m diads in i-PP, all racemic or r diads in s-PP, and a more or less random distribution of both r and m diads in a-PP. To realize the potential of NMR spectroscopy to determine the microstructures of polymers, we must establish the connections between microstructural features and the corresponding chemical shifts of the nuclei residing there. The remarkable sensitivity of the resonance frequency of a nucleus to its electronic environment is at the same time both good and bad news. The good news is that NMR can be used to sensitively probe the microstructures of polymers, but only if we can understand how the microstructure affects the cloud of electrons surrounding each nucleus that is a part of each microstructure. Because it remains very difficult to predict or calculate how the electronic environment of each nucleus in the presence of the applied magnetic field B_0 will respond to changes in local microstructure, the bad news is that *a priori* chemical shift predictions are not generally possible (Ditchfield, 1976; Schastnev and Cheremisin, 1982). Instead, the effects of substituents and local conformation have been successfully utilized to correlate ^{13}C NMR chemical shifts and the microstructures of molecules, including polymers (Duddeck, 1986; Tonelli, 1989).

^{13}C NMR studies of paraffinic hydrocarbons (Spiesecke and Schneider, 1961; Grant and Paul, 1964; Lindeman and Adams, 1971; Dorman, Carhart, and Roberts, 1974; Bovey, 1974) have led to substituent rules useful in the prediction of their ^{13}C chemical shifts $\delta^{13}C$'s. $\delta^{13}C$'s are ordered in terms of the effects produced by substituents attached to the observed carbon at the α, β, and γ positions. These substituent effects are illustrated in Tables 4.1, 4.2, and 4.3,

TABLE 4.1. α-Substituent Effect on $\delta^{13}C$ (Bovey, 1974)

		$\delta^{13}C$ from TMS, ppm	α-Effect, ppm
(a)	$^{o}CH_3$———H	−2.1	—
(b)	$^{o}CH_3$———$^{\alpha}CH_3$	5.9	8.0
(c)	$^{o}CH_2$ $\Big\langle \begin{array}{c} ^{\alpha}CH_3 \\ ^{\alpha}CH_3 \end{array}$	16.1	10.2
(d)	^{o}CH $\Big\langle \begin{array}{c} ^{\alpha}CH_3 \\ ^{\alpha}CH_3 \\ ^{\alpha}CH_3 \end{array}$	25.2	9.1
(e)	^{o}C $\Big\langle \begin{array}{c} ^{\alpha}CH_3 \\ ^{\alpha}CH_3 \\ ^{\alpha}CH_3 \\ ^{\alpha}CH_3 \end{array}$	27.9	2.7

TABLE 4.2. β-Substituent Effect on $\delta^{13}C$ (Bovey, 1974)

		$\delta^{13}C$ from TMS, ppm	β-Effect, ppm
(a)	$^{o}CH_3$———$^{\alpha}CH_3$	5.9	—
(b)	$^{o}CH_3$———$^{\alpha}CH_2$———$^{\beta}CH_3$	15.6	9.7
(c)	$^{o}CH_3$———$^{\alpha}CH_3$ $\Big\langle \begin{array}{c} ^{\beta}CH_3 \\ ^{\beta}CH_3 \end{array}$	24.3	8.7
(d)	$^{o}CH_3$———$^{\alpha}C$ $\Big\langle \begin{array}{c} ^{\beta}CH_3 \\ ^{\beta}CH_3 \\ ^{\beta}CH_3 \end{array}$	31.5	7.2

TABLE 4.3. γ-Substituent Effect on $\delta^{13}C$ (Bovey, 1974)

	$\delta^{13}C$ from TMS, ppm	γ-Effect, ppm
(a) $^{0}CH_3$——$^{\alpha}CH_2$——$^{\beta}CH_3$	15.6	—
(b) $^{0}CH_3$——$^{\alpha}CH_2$——$^{\beta}CH_2$——$^{\gamma}CH_3$	13.2	−2.4
(c) $^{0}CH_3$——$^{\alpha}CH_2$——$^{\beta}CH$ < $^{\gamma}CH_3$ $^{\gamma}CH_3$	11.5	−1.7
(d) $^{0}CH_3$——$^{\alpha}CH_2$——$^{\beta}C$ < $^{\gamma}CH_3$ — $^{\gamma}CH_3$ — $^{\gamma}CH_3$	8.7	−2.8
(e) $^{\alpha}CH_3$——$^{0}CH_2$——$^{\alpha}CH_2$——$^{\beta}CH_3$	25.0	
(f) $^{\alpha}CH_3$——$^{0}CH_2$——$^{\alpha}CH_2$——$^{\beta}CH_2$——$^{\gamma}CH_3$	22.6	−2.4
(g) $^{\alpha}CH_3$——$^{0}CH_2$——$^{\alpha}CH_2$——$^{\beta}CH$ < $^{\gamma}CH_3$ $^{\gamma}CH_3$	20.7	−1.9
(h) $^{\alpha}CH_3$——$^{0}CH_2$——$^{\alpha}CH_2$——$^{\beta}C$ < $^{\gamma}CH_3$ — $^{\gamma}CH_3$ — $^{\gamma}CH_3$	18.8	−1.9

where it is seen that carbon subsituents α and β to an observed carbon result in a deshielding of approximately +9 ppm, while a shielding of approximately −2 ppm is produced by γ carbon substituents, and they have been used to predict the $\delta^{13}C$'s of a wide variety of paraffinic hydrocarbons including highly branched compounds. Let us apply these substituent effects to PP as indicated below.

$$1\alpha, 2\beta, 2\gamma$$

$$\begin{array}{cccc} C & C & C & C \\ | & | & | & | \\ -C-C-C-C-C-C-C-C-C- \end{array}$$

$$2\alpha, 4\beta, 2\gamma \quad 3\alpha, 2\beta, 4\gamma$$

By subtracting 1α, 2β, and 2γ effects from each carbon type, $\delta(CH_3) = 0.0$ ppm, $\delta(CH) = 2\alpha + 2\gamma = 2(-2) + 2(9) = 14$ ppm, and $\delta(CH_2) = 1\alpha + 2\beta = 9 + 2(9) = 27$ ppm. Thus we would expect to observe the CH_3 carbons in PP to resonate most upfield (nearest TMS), the CH carbons to be 14 ppm downfield from the CH_3 carbons, and the CH_2 carbons to be even further downfield 27 ppm from CH_3 carbons and $27-14 = 13$ ppm from the CH carbons. We can see from Figure 4.3 that experimentally the CH_3 carbons resonate most upfield, the CH carbons resonate 8–9 ppm further downfield, and the CH_2 carbons resonate the most downfield, 25–26 ppm from CH_3's and 17–18 ppm from the CH's in agreement with the $\delta^{13}C$'s estimated with the substituent effects. Thus it appears that we can successfully employ the α, β, and γ substituent effects to assign the resonances in the ^{13}C NMR spectra of PP as to carbon type, but how do we assign the multiple resonances observed for each carbon type in a-PP or understand why the resonance frequencies for each carbon type are different for the stereoregular i-PP and s-PP samples?

It is apparent that the stereosequence or tacticity of PP affects its ^{13}C resonance frequencies. However, the numbers and types of α, β, and γ substituents attached to each carbon type in PP are independent of stereosequence. The resolution of this puzzling sensitivity to stereosequence can be achieved with the help of Figure 4.4, where the Newman projections of the trans and gauche conformations about the central bond connecting the observed and γ carbons in an alkane fragment are presented. Note that in the *gauche* conformation $C°$ and C^γ approach each other much more closely (3 Å) than in the *trans* conformation. It has been demonstrated that a γ substituent must approach the observed carbon closely before it can exert its shielding effect (Grant and Cheney, 1967; Li and Chesnut, 1985; Seidman and Maciel, 1977). Consequently, a carbon nucleus in a *trans* conformation with its γ substituent remains unshielded, while in the *gauche* conformation it does become shielded by the γ substituent. In other words, the γ-substituent effect on $\delta^{13}C$'s is conformationally sensitive.

This can be demonstrated with the help of Figure 4.5, where the methyl carbon chemical shifts of three 1-substituted propanes are presented and referred to the resonant frequency of the methyl carbon in propane, which does not have a γ substituent (Tonelli, 1989). Also presented here are the probabilities of finding, or the population of *gauche* bond conformers. When the shifts in the methyl carbon reso-

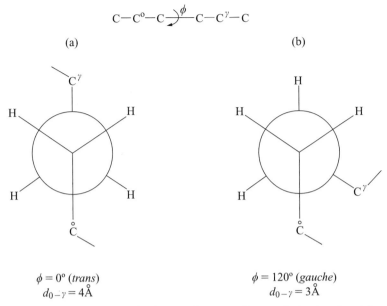

$\phi = 0°$ (*trans*)
$d_{0-\gamma} = 4\text{Å}$

$\phi = 120°$ (*gauche*)
$d_{0-\gamma} = 3\text{Å}$

Figure 4.4. Newman projections of an *n*-alkane chain in the (a) *trans* ($\phi = 0°$) and (b) *gauche* ($\phi = 120°$) conformations.

nances observed from propane are divided by the *gauche* bond populations, we obtain an estimate of the shielding produced by these substituents when in a *gauche* arrangement with the observed methyl; $\gamma_{c,c} = -5.2$ ppm, $\gamma_{c,0} = -7.2$ ppm, and $\gamma_{c,cl} = -6.8$ ppm. We now see that the shielding at a nucleus produced by a γ substituent can be comparable in magnitude (-5 to -7 ppm) to the deshielding ($+9$ ppm) caused by the more proximal α and β substituents. More important, however, is the conformational dependence of the γ-substituent effect on δ^{13}C's. Any variation in the microstructure of a polymer which affects its local conformation can be expected to be reflected in its ^{13}C chemical shifts via the γ-*gauche* effect.

In the next chapter, which discusses the conformational properties of polymers, the dependence of vinyl polymer conformations on their stereosequence will be demonstrated. If here we accept the stereosequence-dependence of vinyl polymer conformations, then we can understand the stereosequence-dependent δ^{13}C's observed in PP. Figure 4.6(a) presents an expansion of the methyl carbon region of the ^{13}C NMR spectrum of a-PP (see Figure 4.3), where nearly 20

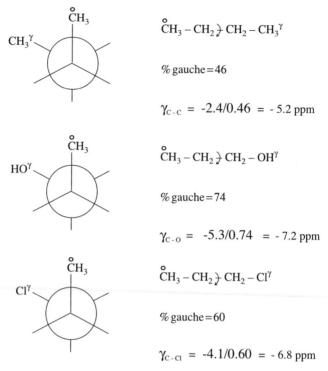

$\overset{o}{C}H_3 - CH_2 \} CH_2 - CH_3{}^{\gamma}$

% gauche = 46

$\gamma_{C-C} = -2.4/0.46 = -5.2\ ppm$

$\overset{o}{C}H_3 - CH_2 \} CH_2 - OH^{\gamma}$

% gauche = 74

$\gamma_{C-O} = -5.3/0.74 = -7.2\ ppm$

$\overset{o}{C}H_3 - CH_2 \} CH_2 - Cl^{\gamma}$

% gauche = 60

$\gamma_{C-Cl} = -4.1/0.60 = -6.8\ ppm$

Figure 4.5. Derivation of the *γ-gauche* shielding produced by the *γ*-substituents, C, OH, and Cl (see text).

different resonances are observed (Tonelli and Schilling, 1981). The number of observed resonances indicates the sensitivity to stereosequence. If the carbon resonance frequency were only sensitive to the stereosequences of the diads on either side of the methine carbon to which it is attached, then we would expect a maximum of three resonances corresponding to mm, mr = rm, and rr triad stereosequences. If the sensitivity to stereosequences extended to nearest-neighbor diads as well, then we could expect to observe a maximum of 10 resonances corresponding to the mmmm, mmmr, mmrm, mmrr, mrrm, mrmr, rmmr, rrmr, rrrm, and rrrr pentad stereosequences. Because well over 20 resonances are observed [see Figure 4.6(a)], the resonance frequency of the methyl carbon nuclei in a-PP must be sensitive to heptad stereosequences, such as that depicted in Figure 4.7, of which there are 36 (*mmmmmm, mmmmmr, mmmmrm,*

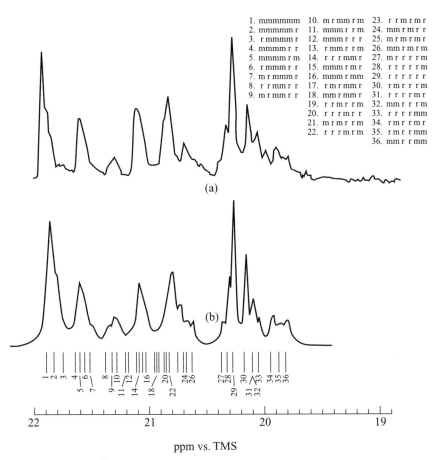

1. mmmmmm	10. m r mm r m	23. r r m r m r
2. mmmmm r	11. mmm r r m	24. mm r m r r
3. r mmmm r	12. mmm r r r	25. m r m r m r
4. mmmm r r	13. r mm r r m	26. mm r m r m
5. mmmm r m	14. r r r mm r	27. m r r r r m
6. r mmm r r	15. mmm r m r	28. r r r r r m
7. m r mmm r	16. mmm r mm	29. r r r r r
8. r r mm r r	17. r m r mm r	30. r m r r r m
9. m r mm r r	18. mm r mm r	31. r r r r m r
	19. r r m r r m	32. mm r r r m
	20. r r r m r r	33. r r r r mm
	21. m r m r r m	34. r m r r m r
	22. r r r m r m	35. r m r r mm
		36. mm r r r mm

(a)

(b)

22   21   20   19

ppm vs. TMS

Figure 4.6. (a) ^{13}C NMR spectrum at 90.52 MHz of the methyl carbon region in atactic PP in 20% w/v n-heptane solution at 67°C. (b) Simulated spectrum obtained from calculated chemical shifts, as represented by the line spectrum below, assuming Lorentzian peaks of <0.1 ppm width at half height. [Reprinted with permission from Tonelli and Schilling (1981).]

mmmrmm, mmmmmr, mmmmrrm, mmrrmm, mmmmrmr, mmrmmr, mrmmmr, rmmmmr, mmrrrm, mmrrmr, mrrmrm, mrrmmr, rrmmmr, mrmmrr, +17 more with *m* and *r* exchanged + *mmmmrr* + *mrmrmr*).

In the heptad stereosequence of a-PP presented in Figure 4.7 the central methyl carbon (CH_3^*) is potentially γ-*gauche* to the neighboring methine carbons C_α, depending on the conformations of the cen-

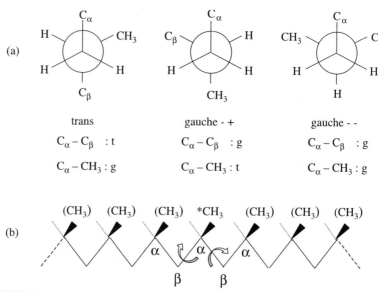

Figure 4.7. (a) Conformations of a four-carbon fragment of a polypropylene chain; (b) heptad of polypropylene chain: observed methyl is marked by *.

tral bonds $(C_\alpha-C_\beta)$ connecting them. The Newman diagram depicting rotation about the left $C_\alpha-C_\beta$ bond makes clear that CH_3^* and C_α are γ-*gauche* when this bond is in the *trans* and *gauche*− conformations. From a knowledge of the *trans* and *gauche* probabilities of both adjoining $C_\alpha-C_\beta$ bonds for each of the 36 heptad stereosequences (see Chapter 5), we may calculate the chemical shifts expected for CH_3^* in each heptad stereosequence. When the resultant probabilities of finding CH_3^* in a *gauche* arrangement with its γ-substituents (C_α) are multiplied by the shielding produced by this arrangement $(\gamma_{c,c} = -5$ ppm), the stick spectrum at the bottom of Figure 4.6(b) is obtained (Tonelli and Schilling, 1981).

Because the γ-*gauche*-effect method of calculating ^{13}C chemical shifts only leads to the prediction of stereosequence-dependent relative chemical shifts, we are free in the comparison with observed spectra to translate the calculated shifts uniformly as a group to obtain the best agreement with the observed $\delta^{13}C$'s. This has been done in Figure 4.6(b), and the agreement between observed and calculated methyl $\delta^{13}C$'s has been used to make the stereosequence assignments indicated there. Measurement of the intensities of the assigned resonances permits a quantative assessment of how much of each stereosequence is present in this sample of a-PP. This application of the conformationally sensitive *γ-gauche* method of assigning the stereo-sequence-dependent ^{13}C NMR spectra of vinyl polymers points up the highly sensitive, long-range microstructural dependence of their ^{13}C NMR spectra and our ability to understand that this dependence results from the extreme sensitivity of the local conformation to the same microstructural feature, the vinyl polymer stereosequences.

DISCUSSION QUESTIONS

1. Is polymer microstructure of concern, or of more concern, when trying to understand the difference in behavior of polystyrene and polyethylene or between ethane and polyethylene and why?

2. Would you expect an atactic or an isotactic vinyl polymer or both to be able to crystallize and why?

3. If the conditions (temperature, pressure, solvent, catalyst, etc.) under which a vinyl polymer is polymerized are altered and the resulting samples of polymer produced exhibit different physical properties, what is likely occurring during their polymerizations?

*4. In Figure 4.3, which presents the ^{13}C NMR spectra of i, s-, and a-PP, we can see that the CH_2 carbons resonate at ~ 46 ppm, the CH carbons at ~ 28 ppm, and the CH_3 carbons at ~ 20 ppm. However, in certain PP samples, minor ^{13}C resonance peaks at 37, 31, and 15 ppm may appear. Using substituent effects on ^{13}C chemical shifts ($\alpha = \beta = 9$ ppm and $\gamma = -2.5$ ppm), suggest what type of PP microstructure(s) may be producing these minor resonances.

*5. In the ^{13}C NMR spectrum of *n*-dodecane, six resonances are observed. Is this consistent with expectations based on α, β, and γ substituent effects? If not, what might be the reason for observing six resonances?

*6. Polyethylenes are often designated as low density (LD), high density (HD), and linear, low density (LLD) samples or products. LDPE contains long branches produced by chain transfer to polymer, HDPE is linear without branches, and LLDPE has a small number of short branches achieved through copolymerization with comonomers like propylene, 1-butene, and 1-hexene, for example. Why are LDPE and LLDPE lower density materials than HDPE? Also, based on α-, β-, and γ-substituent effects, what would you expect the ^{13}C NMR spectra of these three PE's to look like?

REFERENCES

Bovey, F. A. (1974), in *Proceedings of the International Symposium on Macromolecules, Rio de Janeiro, July 26–31, 1974*, E. B. Mano, Ed., Elsevier, New York, p. 169.

Ditchfield, R. (1976), *Nucl. Magn. Reson.*, **5**, 1.

Dorman, D. E., Carhart, R. E., and Roberts, J. D. (1974), private communications cited in Bovey (1974).

Duddeck, H. (1986), in *Topics in Stereochemistry*, Eliel, E. L, Witen, S. H., and Allinger, N. L., Eds., Wiley-Interscience, New York, p. 219.

Ferry, J. D. (1970), *Viscoelastic Properties of Polymers*, 2nd ed., Wiley, New York.

Grant, D. M., and Cheney, B. V. (1967), *J. Am. Chem. Soc.*, **89**, 5315.

Grant, D. M., and Paul, E. G. (1964), *J. Am. Chem. Soc.*, **86**, 2984.

Hirooka, M., and Kato, T. (1974), *J. Polym. Sci. Polym. Lett. Ed.*, **12**, 31.

Li, S., and Chesnut, D. B. (1985), *Magn. Reson. Chem.*, **23**, 625.

Lindeman, L. P., and Adams, J. Q. (1971), *Anal. Chem.*, **43**, 1245.

Osgen, M., and Price, C. C. (1959), *J. Polym. Sci.*, **34**, 153.

Price, C. C., and Osgen, M. (1956), *J. Am. Chem. Soc.*, **78**, 4787.

Pruitt, M. E., and Baggett, J. M. (1955), U.S. Patent #2,706,181.

Schastnev, P. V., and Cheremisin, A. A. (1982), *J. Struct. Chem.*, **23**, 44.

Seidman, K., and Maciel, G. E. (1977), *J. Am. Chem. Soc.*, **99**, 659.

Spiesecke, H., and Schneider, W. G. (1961), *J. Chem. Phys.*, **35**, 722.

Tonelli, A. E. (1975), *Macromolecules*, **8**, 544.

Tonelli, A. E. (1986), in *Cyclic Polymers*, Semlyen, A. J., Ed., Elsevier, London, Chapter 8.

Tonelli, A. E. (1989), *NMR Spectroscopy and Polymer Microstructure: The Conformational Connection*, VCH, New York.

Tonelli, A. E., and Schilling, F. C. (1981), *Accts. Chem. Res.*, **14**, 233.

Tsurata, T. (1967), in *The Stereochemistry of Macromolecules*, Ketley, A. D., Ed., Marcel Dekker, New York.

Ward, I. M. (1985), *Adv. Polym. Sci.*, **70**, 1.

CHAPTER 5

THE CONFORMATIONAL CHARACTERISTICS OF POLYMERS

The purpose of this chapter is to describe the conformational characteristics of polymers, which govern their overall sizes and shapes and thus their responses to environmental stimuli. The conformation of a polymer molecule describes the geometrical arrangement of the atoms in its chain that may be achieved principally through rotations about the chemical bonds composing its backbone. It is generally justified to ignore the additional conformational freedom resulting from stretching or compressing the backbone bonds or opening or closing the backbone valence angles, because not only are the energies resisting these deformations much higher than the energies required for rotation about the backbone bonds, but changes in the overall size or shape of a polymer chain produced by backbone bond rotations far exceed those produced by bond stretching and/or bond angle bending. This is clearly illustrated in Figure 5.1, where the terminal units of a polymer chain fragment are brought close together in space as a consequence of a single backbone bond rotation.

ROTATIONAL ISOMERIC STATE MODEL

Though the energy necessary to rotate about a sp^3-sp^3, C–C single bond in a polymer backbone is generally much less than is required to

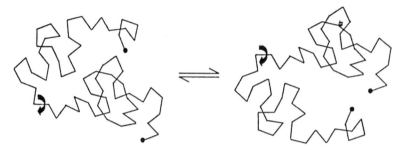

Figure 5.1. Two polymer chain fragments whose conformations differ only about one of their backbone bonds.

stretch or compress the C–C bond lengths or bend (open or close) the $\begin{smallmatrix} C \\ / \ \backslash \\ C \ \ C \end{smallmatrix}$ valence angles, it is not completely negligible. Spectroscopic studies of low-molecular-weight compounds (Mizushima, 1954; Wilson, 1959, 1962; Herschback, 1963) and electron diffraction (Bonham and Bartell, 1959; Kuchitsu, 1959; Bartell and Kohl, 1963) have clarified the nature of the potentials hindering rotations about chemical bonds. Rotation about the sp^3-sp^3, C–C bond in ethane (and all other alkanes including polyethylene) is threefold, with energy minima corresponding to the staggering of the methyl hydrogen atoms. When the methyl hydrogen atoms are eclipsed, repulsion of the bonding electrons in each pair of apposed C–H bonds is at a maximum, leading to rotation away from the eclipsed to a staggered conformation, which relieves the electronic repulsion. The rotation potential $E(\phi_2)$ about the central bond in butane is represented in Figure 5.2. As for ethane, the rotation potential is threefold but no longer symmetrical, owing to the nonbonded interactions of the terminal methyl groups in the *gauche* rotational states at $\phi_2 = \pm120°$. (Rotation angles are taken as $0°$ in the planar zigzag or *trans* conformation and assume positive values for right-handed rotations.) The barrier separating the *gauche* ($\phi_2 = \pm120°$) and *trans* ($\phi_2 = 0°$) rotational states in *n*-butane is ~3.5 kcal/mol (14.6 kJ/mol), which considerably exceeds RT at ordinary temperatures, where R = gas constant. Since RT is a measure of the available thermal energy, the distribution of rotation angles over the 2π range is expected to be nonuniform at equilibrium. With values of $\phi_2 = 0°$, $\pm120°$ being heavily preferred over $\phi_2 = 180°$, $\pm60°$.

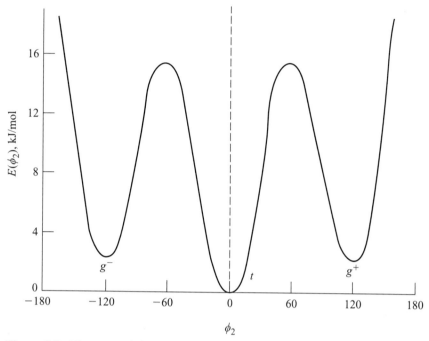

Figure 5.2. The potential energy $E(\phi_2)$ of rotation ϕ_2 about the central C–C bond in butane CH_3–CH_2⤳$^{\phi_2}CH_2$–CH_3. Both methyl groups are fixed in the staggered positions. To convert J to cal, divide by 4.184 (Flory, 1969; Tonelli, 1986).

A realistic representation of the internal rotational state of *n*-butane is afforded by considering each molecule to be confined to small oscillations (e.g., ±20°) about each of its three energy minima $\phi_2 = 0°$, ±120°. Though the barriers separating the minimum energy conformations are not large enough to prevent rapid transitions between them, they do assure that a negligible population of *n*-butane molecules have $\phi_2 = 180°$, ±60°. The discrete nature of the rotational potential about the backbone bonds in a polymer chain has been recognized (Volkenstein, 1963; Birshtein and Ptitsyn, 1964) in the rotational isomeric state (RIS) approximation, where each backbone bond in the polymer chain is assumed to be confined to any one of a small number of discrete rotational states. These states are generally selected to coincide with the potential minima (see Figure 5.2). For polymers with backbone bonds whose rotational barriers do not exceed RT, the

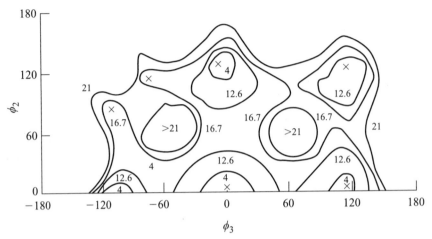

Figure 5.3. Conformational energy $E(\phi_2, \phi_3)$ map for *n*-pentane with $\phi_1 = \phi_4 = 0°$. Minima are indicated by an x, and contours are shown in kJ/mol. To convert J to cal, divide by 4.184.

n-Pentane conformations $(CH_3-CH_2\overset{\phi_2}{\rightleftharpoons}CH_2\overset{\phi_3}{\rightleftharpoons}CH_2-CH_3)$: (a), *trans, trans* $(\phi_2 = \phi_3 = 0°)$; (b), *gauche*$^+$, *gauche*$^+$ $(\phi_2 = \phi_3 = 120°)$; and (c), *gauche*$^+$, *gauche*$^-$ $(\phi_2 = 120°, \phi_3 = -120°)$.

RIS model may be used to approximate the nearly continuous rotation potential by a discrete sum of rotational states (Mansfield, 1983).

Not only is each backbone bond in a polymer chain usually restricted to reside in a few discrete rotational states, but the probability of occurrence of any rotational state about a given bond depends on the roational states of its nearest-neighbor backbone bonds (Volkenstein, 1963; Birshstein and Ptitsyn, 1964; Flory, 1969). This rotational state interdependence is easily demonstrated by the conformations of *n*-pentane as depicted in Figure 5.3.

Severe repulsive steric interaction between the terminal methyls occurs in the g^+g^- (and g^-g^+) conformation of Figure 5.3(c), whereas in the g^+g^+ (and g^-g^-) conformation of Figure 5.3(b) the methyls are sufficiently separated (by $\sim 3.6\,\text{Å}$) to suggest an approximately neutral (neither attractive nor repulsive) interaction. The energy of rotational state ϕ_2 of bond 2 evidently depends on the rotational state ϕ_3 of bond 3 in n-pentane. When ϕ_2 is g^-, ϕ_3 prefers to be t (trans) or g^- rather than g^+, owing to the severe interaction between methyl groups in the $(\phi_2, \phi_3) = g^-g^+$ conformation.

In most polymer chains, as in n-pentane, the relative probabilities or populations of the rotational states of a given bond depend on the rotational states of nearest-neighbor bonds. Unlike the energies of the rotational states about a given bond, the locations of the rotational states are unaffected by the rotational states of neighboring bonds. For most polymer chains, this assumption seems justified, and for those polymers where both the locations and energies of rotational states are interdependent, additional rotational states can be incorporated into the RIS conformational model (Flory, 1969; Mattice and Suter, 1994). Nonbonded interactions between atoms and/or groups of atoms separated by four bonds are the source of this nearest-neighbor interdependence of polymer chain conformations (see n-pentane in Figure 5.3, for example). Consequently, most polymer chains can be treated as one-dimensional (linear chains), statistical mechanical systems composed of nearest-neighbor-dependent elements. Later we will see that such systems are conveniently treated by the mathematical methods developed (Kramers and Wannier, 1941; Newell and Montroll, 1953) to treat the one-dimensional Ising model of ferromagnetism (Ising, 1925).

For the moment let us consider the polyethyene (PE) chain, as depicted in Figure 5.4, where both one-bond (t, g^+, g^-) and two-bond (tt, g^+g^+, g^+g^-) rotational conformations are illustrated. A comparison with the n-pentane conformers of Figure 5.3 makes apparent that the two-bond, nearest-neighbor-dependent conformations in PE and n-pentane result in very similar interactions, which as a first approximation we treat as identical. We may use the conformational energy maps for n-butane (Figure 5.2) and n-pentane (Figure 5.3) to determine the population of each of the two-bond, nearest-neighbor-dependent conformations in PE. From Figure 5.2 we can estimate that the energy of a *gauche* conformation is ~ 500 cal/mol higher than a *trans* conformation, or $E_{g\pm} = 500$ cal/mol and $E_t = 0.0$ cal/mol. Comparison of the energies of the n-pentane conformers in Figure 5.3

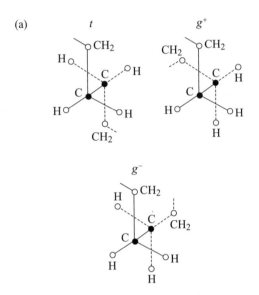

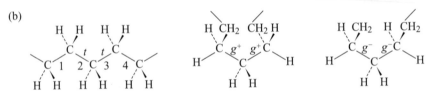

Figure 5.4. (a) One- and (b) two-bond conformations in a PE chain.

leads to the following two-bond conformational energies: $E_{tt} = 0.0$, $E_{tg\pm} = E_{g\pm t} = E_{g\pm} = 500$ cal/mol, $E_{g\pm g\pm} = 2E_{g\pm} = 1000$ cal/mol, and $E_{g\pm g\mp} = 2E_{g\pm} + E_\omega = 1000 + 2000 = 3000$ cal/mol. E_ω is the energy of interaction between the terminal methyl groups in the $g\pm g\mp$ n-pentane conformers. If each two-bond PE conformation is weighted exponentially according to its energy, then their populations or probabilities can be obtained (Flory, 1969) from $P(\phi_i, \phi_{i+1}) = \exp-[E(\phi_i, \phi_{i+1})/RT]/\{\sum_{\phi_i,\phi_{i+1}} \exp - [E(\phi_i, \phi_{i+1})/RT]\}$, which leads at $T = 25°C$ to $P_{tt} = 0.322$, $P_{tg\pm} = P_{g\pm t} = 0.139$, $P_{g\pm g\pm} = 0.060$, and $P_{g\pm g\mp} = 0.001$.

These two-bond conformer populations are only strictly applicable to n-pentane and are approximate for PE, because the rotational state

ϕ_i of bond i in PE depends on the conformations of both the preceding (ϕ_{i-1}) and succeeding (ϕ_{i+1}) bonds. Later in this chapter we will indicate how the two-bond conformer populations can be appropriately calculated for PE. Nevertheless, summation over the two-bond conformer populations above leads to the conclusion that in *n*-pentane and PE each backbone bond is $\sim 60\%$ *trans* and 40% *gauche+* or *gauche−*.

POLYMER MICROSTRUCTURES AND CONFORMATIONS

Of course the population of polymer backbone conformations depends on their detailed microstructures. In Figure 5.5 the conformational energy map for a meso (m) diad of poly(2-vinylpyridine) (P2VP) is presented. The conformational energy map for a racemic (r) P2VP diad, where the second H and pyridine ring were interchanged, is given in Figure 5.6. Note that in both P2VP diad conformational energy maps, low-energy $[E(\phi_1, \phi_2)]$ regions are limited to only two values of each rotation angle. Thus, ϕ_1 is restricted to t and g^+ rotational states in both diads, while ϕ_2 is confined to t and g^- conformations in the *m* diad and to t and g^+ conformations in the racemic diad. Unlike PE, the backbone bonds in P2VP are restricted to just two and not all three staggered rotational states. The elimination of one of the staggered conformations about each of the P2VP backbone bonds is understandable based on the Newman projections presented in Figure 5.7. The bulky CH_2, $C^\alpha H$, and Pyr groups are in close proximity when $\phi_1 = g^-$ and when $\phi_2 = g^+$, so each of these staggered backbone conformations is precluded for P2VP.

The populations of two-bond, nearest-neighbor-dependent conformations for P2VP $m(r)$ diads obtained from their energy maps are $P_{tt} = 0.074\ (0.248)$, $P_{tg+} = P_{g+t} = 0.325\ (0.045)$, and $P_{g+g-} = 0.276(m)$ and $P_{g+g+} = 0.662(r)$. Not only are the populations of bond conformations in P2VP vastly divergent from those in PE, but they also differ substantially between the *m* and *r* diads of the P2VP chain. Clearly the microstructures of polymers can have a significant impact upon their local conformational characteristics, which we will subsequently demonstrate to be a major influence on the physical properties they manifest.

Recall in the previous chapter that the methyl carbon region of the

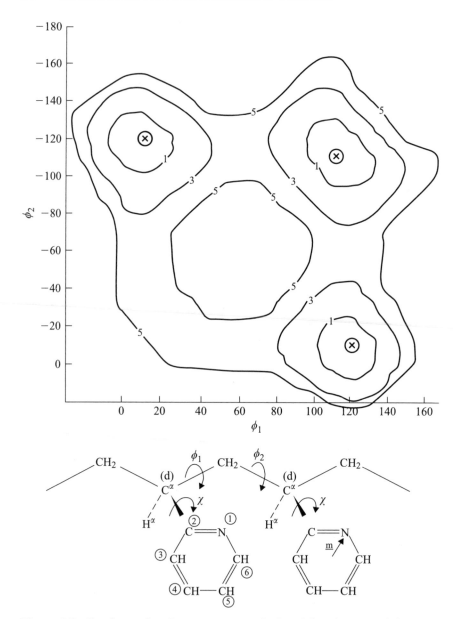

Figure 5.5. Conformational energy map calculated for the *meso* (*m*) P2VP diad drawn below in the *tt* ($\phi_1 = \phi_2 = 0°$) conformation. The pyridine side group was fixed in the conformation where C2–C3 and C^α–H^α are eclipsed. X indicates locations of the lowest energy conformations, and the contours are drawn in kcal/mol relative to X. (After Tonelli, 1985.)

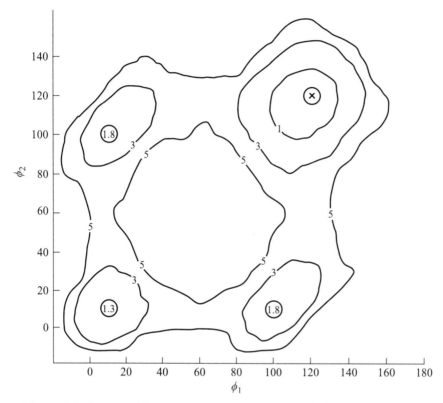

Figure 5.6. Same as Figure 5.5 except replace meso (*m*) with racemic (*r*).

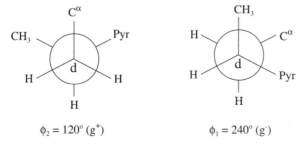

$\phi_2 = 120°$ (g$^+$) $\phi_1 = 240°$ (g$^-$)

Figure 5.7. Newman projection along the CH_2–C^xH and C^xH–CH_2 bonds in P2VP.

TABLE 5.1. Calculated Probability of Finding
Bond ϕ in the *trans* Conformation

$$
\begin{array}{cccccccc}
 & \text{C} & m,r & \text{C} & m,r & \text{C} & m,r & \text{C} & m,r & \text{C} \\
 & | & & | & & | & & | & & | \\
-\text{C}-\text{C}- & \text{C}- & \text{C}- & \text{C}- & \overset{\phi}{-} & \text{C}- & \text{C}- & \text{C}- & \text{C}-
\end{array}
$$

Pentad Stereosequence	$P(\phi = t)$
mrmr	.440
rrmr	.472
mmmm	.523
rmmr	.539
rmmm	.582
rrrr	.635
mrrm	.685
rrrm	.712
mmrr	.742
rmrm	.763
mmrm	.792

^{13}C NMR spectrum of atactic polypropylene (a-PP) was assigned by quantiative application of the *γ-gauche* effect method of estimating ^{13}C chemical shifts. The critical information needed to estimate the stereosequence dependence of methyl carbon chemical shifts was the conformational populations of the C–C backbone bonds adjoined to the asymmetric carbon bonded to the methyl group. In Table 5.1 the probabilties of finding one such C–C bond in the trans conformation (P_t) are presented for the various pentad stereosequences of a-PP (Tonelli, 1989). Note that among the different a-PP pentads, ΔP_t is 0.35; and, because these bond conformer populations must be multiplied by a γ-effect $= -5$ ppm to obtain the methyl carbon chemical shifts, it is not surprising that the ^{13}C chemical shifts observed for the various stereosequences in a-PP are spread over more than a 2 ppm range. Now we can readily appreciate how a knowledge of the conformational characteristics of PP, as embodied in its RIS model, explains the stereosequence-dependent chemical shifts observed in the ^{13}C NMR spectra of stereoirregular a-PP samples, which depend on the local conformations of the a-PP chains.

POLYMER CONFORMATIONS AND SIZES

Because the sizes and shapes available to polymer chains influence their physical properties, we must be able to connect them to their microstructures in order to begin to establish structure–property relations for polymers. A parameter useful for characterizing the size of a polymer chain is its end-to-end distance **r**. A schematic representation is given in Figure 5.8 of a polymer with an all-carbon backbone consisting of n bonds and $n + 1$ atoms. The backbone conformation is determined by the disposition of the set of bond vectors l_1, l_2, l_3, which depends of course on the rotational state of each bond. The vector r connecting the ends of the chain C_0 and C_n is $r = \sum_i l_i$, and its magnitude is simply **r** the distance between the chain ends. Consequently, $\mathbf{r}^2 = r \cdot r = \sum_{i,j} l_i \cdot l_j = \sum_i l_i^2 + 2 \sum_{0 < i < j \le n} l_i \cdot l_j$, the factor of two resulting from the equivalence of $l_i \cdot l_j$ and $l_j \cdot l_i$.

The ability of a polymer to rotate about its chemical bonds makes the number of conformations it may adopt extremely large. As an example, we have seen in the development of RIS conformational models for PE and P2VP that each of their backbone bonds may adopt either three or two conformations, respectively. Thus for 10,000 bond PE (P2VP) chains a total of $3^{10,000}$ ($2^{10,000}$) are available. Complete enumeration of each conformation is a hopeless task, which may, however, be avoided by considering appropriate averages over the totality of available conformations. The mean-square end-to-end distance $\langle \mathbf{r}^2 \rangle$ of a polymer is one such measure of the size of a polymer chain averaged over its multitude of available conformations. $\langle \mathbf{r}^2 \rangle_0 = \sum \langle l_i^2 \rangle + 2 \sum \langle l_i \cdot l_j \rangle$, where $0 < i < j \le n$, and if all bond

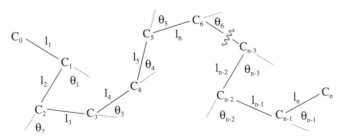

Figure 5.8. Schematic representation of an all-carbon polymer backbone.

lengths l are equal, or if l is identified as the average squared bond length, then $\langle r^2 \rangle_0 = nl^2 + 2 \sum_{0 < i < j \leq n} \langle l_i \cdot l_j \rangle$. The subscript 0 denotes the unperturbed state of the polymer chain, where interactions between chain segments i, j separated by large $|i - j|$ are ignored and all conformations permitted by its RIS model are adopted (see below).

Because of the immense number of conformations available to a typical polymer chain, their direct enumeration to obtain $\langle l_i \cdot l_j \rangle$, which is necessary to evaluate $\langle r^2 \rangle_0$, is often avoided by adoption of model or hypothetical chains. The simplest is the random-flight or freely jointed chain (Rayleigh, 1919; Chandrasekhar, 1943), whose neighboring bonds are completely uncorrelated; that is, all directions of a given bond are equally likely and independent of the neighboring bonds. If the bonds of the polymer in Figure 5.8 are fixed in length and connected to each other at any angle with equal probability, and if rotation about each bond is unrestricted, the polymer chain is freely jointed. For the uncorrelated bonds of a freely jointed chain, we have $\langle l_i \cdot l_j \rangle = 0$, for $i \neq j$ and $\langle r^2 \rangle_0 = nl^2$.

If the valence angles $(\pi - \theta)$ formed by the junctions of neighboring bonds in the freely jointed chain are held fixed, the resulting model chain is more realistic and is termed *freely rotating*. The projection of bond $i + 1$ on i is $l \cos \theta$, and does not average to 0, as in the case of a freely jointed chain (all transverse projections of $i + 1$ on i do, however, average to 0). Thus, $\langle l_i \cdot l_{i+k} \rangle = l^2 (\cos \theta)^k$ and $\langle r^2 \rangle_0 = nl^2 + 2l^2 \sum_{0 < i < j \leq n} (\cos \theta)^{j-i}$, which, after combination of terms and for long chains (large n), becomes $\langle r^2 \rangle_0 = [(1 + \cos \theta)/(1 - \cos \theta)]nl^2$ (Eyring, 1932; Kuhn, 1934; Wall, 1943).

A further step toward reality may be taken in terms of polymer chain models by assuming that the rotation about bonds in the freely rotating chain is no longer free, but restricted. Although restricted, it is assumed to be uncorrelated; that is, the energy or potential $E(\phi_i)$ affecting rotation about bond i depends only on the rotation angle ϕ_i about this bond. This model is called the independently rotating chain. For symmetric chains like PE (no asymmetric centers in the backbone) we have $E(\phi_i) = E(-\phi_i)$ (see Figure 5.3) and $\langle \sin \phi_i \rangle = 0.0$, and for long chains we have $\langle r^2 \rangle_0 = [(1 + \cos \theta)/(1 - \cos \theta)][(1 + \langle \cos \phi \rangle)/(1 - \langle \cos \phi \rangle)]nl^2$ (Oka, 1942). For free rotation, $\langle \cos \phi \rangle = 0$, and we regain the expression for $\langle r^2 \rangle_0$ of the freely rotating chain.

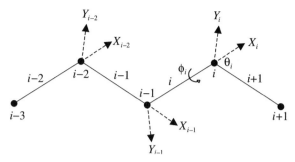

Figure 5.9. A schematic drawing of a polymer-chain backbone illustrating a numbering scheme, bond-rotation ϕ and valence $(\pi - \theta)$ angles, and the Cartesian reference frames defined along each bond.

For purposes of comparison, we may divide $\langle r^2 \rangle_0$ by the number of bonds n times the square of the bond length to obtain the characteristic ratio $C_n = \langle r^2 \rangle_0 / nl^2$ (Flory, 1969). This permits a comparison of the average size of a polymer to that of the most flexible polymer, the freely jointed chain, where $\langle r^2 \rangle_0 = nl^2$. The three model chains just discussed have characteristic ratios of 1.0 (freely jointed), $[(1 + \cos\theta)/(1 - \cos\theta)] = 2.0$ for tetrahedral bonds (freely rotating), and $[(1 + \cos\theta)/(1 - \cos\theta)][(1 + \langle \cos\phi \rangle)/(1 - \langle \cos\phi \rangle)]$ (independently rotating). Clearly, $\langle r^2 \rangle_0$ is proportional to n and C_n is independent of n for each of the model chains of sufficient length. We will see later that this is also true for the dimensions ($\langle r^2 \rangle_0$) obtained rigorously from the nearest-neighbor-dependent RIS description of polymer conformations.

Let us anticipate the RIS results here and accept that $\langle r^2 \rangle_0 = 160,000 \,\text{Å}^2$ for a 10,000-bond unperturbed PE chain, or that the average end-to-end distance is $(\langle r^2 \rangle_0)^{1/2} = 400 \,\text{Å}$. If this same PE chain were completely extended in the all-*trans*, planar zigzag conformation, then $r = 10,000\{l \sin[(\pi - \theta)/2]\} = 12,700 \,\text{Å}$, which is over a factor of 30 times more extended than the unperturbed PE chain having access to its full multitude of conformations. (Referring to Figure 5.9, check by simple trigonometry that the and-to-end distance of a 10,000 bond PE chain in the all-*trans* conformation is indeed 12,700 Å.) This comparison points out the inherent ability of high-molecular-weight polymer chains to vary their sizes over a broad range leading, for example, to the characteristic highly elastic responses of many polymeric materials.

Now consider the same unperturbed PE chain and estimate the volume that it spans or influences as it adopts its myriad conformations. One such estimate can be made by assuming the volume influenced by the coiling PE chain is described by a sphere whose diameter is $(\langle \mathbf{r}^2 \rangle)_0^{1/2} = 400\,\text{Å}$. The volume V_i of this sphere of influence is $3.3 \times 10^7\,\text{Å}^3$ At the same time, the volume physically occuppied by the C and H atoms of the same PE chain can be estimated by dividing its molecular weight ($14 \times 10,000$) by its [density $\times$ (Avogadro's number], which yields a hard-core volume (V_0) of $2.3 \times 10^5\,\text{Å}^3$ that is more than two orders of magnitude smaller than the volume influenced (V_i) by the coiling PE chain. This observation has important consequences for polymers in solution and in their conformationally disordered bulk states. First, in solutions dilute enough to prevent overlap of polymer coils (generally well below 1 wt%), each polymer chain must influence many solvent molecules. This, as we shall examine in the next chapter, results in very large increases in the viscosities of their solutions produced by very small amounts of dissolved polymer. Second, in their disordered bulk states, each coiling polymer chain must be in contact or entangled with a large number of other polymer chains, leading to very high melt viscosities as discussed in Chapter 7.

RIS MODEL AND POLYMER DIMENSIONS

From this point on, we develop briefly the conformational properties of polymers based on the realistic RIS model, which eliminates the need to introduce any of the artificial model chains discussed previously. This is important, because the chain models (freely jointed, freely rotating, independently rotating, etc.) disregard chemical structure and various elements of chain geometry, the two features that serve to distinguish among different polymers, and cannot therefore be used to develop structure–property relations for polymers. Any of the artificial chain models can, however, generally be utilized to distinguish between the behavior of polymers and all other materials composed of atoms or small molecules. Put another way, to begin to understand polymer chemistry we must treat the conformational properties of polymers in a realistic fashion (RIS model), while less realistic model chains are sufficient to gain an appreciation of the physics of polymers. [In Chapter 7, mention will be made of an example where adoption of an artificial model of a polymer chain (pearl necklace or freely jointed) leads to physical behavior that is *not distinguishible* from the behavior exhibited by materials composed of small molecules.]

If all bond lengths and valence angles are assumed to be fixed, the conformation of an n-bond polymer can be specified by assigning a rotational state to each of the $n-2$ nonterminal bonds. Let v be the number of rotational states, usually three, about each bond. Then there are $v^{(n-2)}$ possible conformations in all. The energy $E\{\phi\}$ of any conformation can be expressed in terms of pairwise, nearest-neighbor-dependent energies $E_i(\phi_{i-1}, \phi_i)$, $E\{\phi\} = \sum_{i=2}^{n-1} E(\phi_{i-1}, \phi_i) = \sum_{i=2}^{n-1} E_{\zeta\eta;i}$, where ζ and η denote the rotational states of bonds $i-1$ and i, respectively.

Statistical weights $\mu_{\zeta\eta}$, or Boltzmann factors, corresponding to the energies $E_{\zeta\eta}$ may be defined as $\mu_{\zeta\eta} = \exp[-E_{\zeta\eta}/RT]$ and expressed in matrix form:

$$\mathbf{U}_i = [\mu_{\zeta\eta;i}] = \begin{bmatrix} \mu_{\alpha\alpha}\mu_{\alpha\beta} & \cdots & \mu_{\alpha v} \\ \mu_{\beta\alpha}\mu_{\beta\beta} & \cdots & \mu_{\beta v} \\ \vdots & \vdots & \vdots \\ \mu_{v\alpha}\mu_{v\beta} & \cdots & \mu_{vv} \end{bmatrix}$$

The rows of the $v \times v$ U_i matrix are indexed with the states (ζ) of bond $i-1$ and the columns with the states (η) of bond i. The statistical weight of a particular chain conformation is then simply $\Omega\{\phi\} = \prod_{i=2}^{n-1} \mu_{\zeta\eta;i}$. Summing this expression over all possible chain conformations leads formally to the conformational partition function

$$Z = \sum_{\{\phi\}} \Omega\{\phi\} = \sum_{\{\phi\}} \left(\prod_{i=2}^{n-1} \mu_{\zeta\eta;i} \right) = \sum_{\{\phi\}} \left(\prod_{i=2}^{n-1} \exp[-E(\phi_{I-1}, \phi_i)/RT)] \right).$$

Application (Flory, 1969) of matrix methods (Kramers and Wannier, 1941) previously developed to treat the Ising ferromagnet (Ising, 1925), also a one-dimensional system with nearest-neighbor-dependent elements, leads to $Z = J^* \left[\prod_{i=2}^{n-1} U_i \right] J$, where J^* and J are the $1 \times v$ and $v \times 1$ row and column vectors

$$J^* = [10\ldots0]; J = \begin{bmatrix} 1 \\ 1 \\ 1 \\ '' \\ '' \\ '' \\ 1 \end{bmatrix}$$

The partition function Z is a distribution function of all chain conformations with each conformation weighted according to the Boltzmann factor of its energy $E\{\phi\}$ (Hill, 1960). Z can be utilized to determine various thermodynamic and conformational properties of isolated polymer chains, provided that their RIS models—that is, the number and energies of their bond rotational states—are known. We now illustrate the importance of establishing the partition function Z for a polymer chain by way of an analysis of the bond conformational populations in n-hexane ($CH_3-CH_2^{\phi_1}\!\!-CH_2^{\phi_2}\!\!-CH_2^{\phi_3}\!\!-CH_2-CH_3$), whose conformations are specified by assigning t, g^+, and g^- rotational states to ϕ_1, ϕ_2, and ϕ_3. Our previous examination of n-butane and n-pentane revealed that each *gauche* rotation incurs an energy of ~ 500 cal/mol and that a $g \pm g \mp$ pair requires an additional 2000 cal/mol. As a result, we would expect the energies of the ttt, $tg \pm t$, $tg \pm g \pm$, and $g \pm g \mp t$ conformers to be 0, 500, 1000, and 3000 cal/mol, respectively. If we define σ as the Boltzmann factor of $E_\sigma = 500$ cal/mol, corresponding to a $g \pm$ conformation, and ω as the Boltzmann factor of $E_\omega = 2000$ cal/mol, corresponding to a $g \pm g \mp$ pair, then the statistical weights of the ttt, $tg \pm t, tg \pm g \pm$, and $g \pm g \mp t$ conformers are 1.0, σ, σ^2, and $\sigma^2 \omega$, respectively. The bond conformational populations obtained in this manner for n-hexane are presented in Table 5.2.

We now repeat the calculation of conformational populations in n-hexane by employing the matrix methods described above. The statistical weight matrices U_1, U_2, and U_3 are given by $U_1 = \begin{vmatrix} 1 & \sigma & \sigma \\ 1 & \sigma & \sigma \\ 1 & \sigma & \sigma \end{vmatrix}$,

TABLE 5.2. Energies, Statistical Weights, and Probabilities of *n*-Hexane Conformations

φ_1	φ_2	φ_3	$E(\varphi_1, \varphi_2, \varphi_3)$, kcal/mole	$\mu(\varphi_1, \varphi_2, \varphi_3)$	$P(\varphi_1, \varphi_2, \varphi_3)$
t	t	t	0.0	1.0	0.189
t	t	$g\pm$	0.5	$\sigma = 0.43$	0.081
$g\pm$	t	t	0.5	$\sigma = 0.43$	0.081
t	$g\pm$	t	0.5	$\sigma = 0.43$	0.081
t	$g\pm$	$g\pm$	1.0	$\sigma^2 = 0.19$	0.036
$g\pm$	$g\pm$	t	1.0	$\sigma^2 = 0.19$	0.036
$g\pm$	t	$g\pm$	1.0	$\sigma^2 = 0.19$	0.036
$g\pm$	t	$g\mp$	1.0	$\sigma^2 = 0.19$	0.036
$g\pm$	$g\pm$	$g\pm$	1.5	$\sigma^3 = 0.08$	0.015
t	$g\pm$	$g\mp$	3.0	$\sigma^2\omega = 0.006$	0.001
$g\pm$	$g\mp$	t	3.0	$\sigma^2\omega = 0.006$	0.001
$g\pm$	$g\pm$	$g\mp$	3.5	$\sigma^3\omega = 0.003$	0.0005
$g\mp$	$g\pm$	$g\pm$	3.5	$\sigma^3\omega = 0.003$	0.0005
$g\pm$	$g\mp$	$g\pm$	5.5	$\sigma^3\omega^2 = 0.0$	0.0

$$z = \sum_{\varphi_1\varphi_2\varphi_3} \mu(\varphi_1, \varphi_2, \varphi_3) = 1 + 6\sigma + 8\sigma^2 + 4\sigma^2\omega + 2\sigma^3 + 4\sigma^3\omega + 2\sigma^3\omega^2 = 5.29$$

and $P(\varphi_1, \varphi_2, \varphi_3) = \mu(\varphi_1, \varphi_2, \varphi_3)/z$

and $U_2 = U_3 = \begin{vmatrix} 1 & \sigma & \sigma \\ 1 & \sigma & \sigma\omega \\ 1 & \sigma\omega & \sigma \end{vmatrix}$. $Z = J^* \left[\prod_{i=2}^{n-1} U_i \right] J = J^*[U_1 U_2 U_3]J = 1 + 6\sigma + 8\sigma^2 + 4\sigma^2\omega + 2\sigma^3 + 4\sigma^3\omega + 2\sigma^3\omega^2$, which not surprisingly is the sum of statistical weights obtained "by hand" in Table 5.2. Now we wish to determine the probability that $\Phi_2 = t$. From Table 5.2 the sum of statistical weights for all conformations with $\Phi_2 = t$ is $1 + 4\sigma + 4\sigma^2$, which, when divided by the sum of the statistical weights for all conformations, yields 0.658. If U_2 is replaced with $U_2' = \begin{vmatrix} 1 & 0 & 0 \\ 1 & 0 & 0 \\ 1 & 0 & 0 \end{vmatrix}$, which reflects fixing $\Phi_2 = t$, then the matrix method gives the probability of $\Phi_2 = t$ as Z'/Z, where in Z' U_2 is replaced by U_2', and $P(\Phi_2 = t) = 0.658$. It is left as an exercise for the reader to use the matrix method to evaluate and check all 27-conformer probabilities given in Table 5.2. (As an example, to calculate $P(g^+tg^-)$ replace the the first and third columns of U_1 with 0's, replace all of the elements of U_2 with 0's except the 2,1 element, and replace all the elements of U_3 with 0's except the 1,3 element.)

We now turn our attention to evaluating polymer properties, which depend on the over all conformation of the polymer and therefore must be averaged over all conformations available to the chain. The mean-square end-to-end distance $\langle r^2 \rangle_0$ is such a property, and $\langle r^2 \rangle_0 = nl^2 + 2 \sum_{0 < i < j \leq n} \langle l_i \cdot l_j \rangle$. Omitting the details (Flory, 1969), we simply mention that the sum of all scalar products between all non-identical backbone bond vectors $l_i \cdot l_j$ $(i \neq j)$ must be averaged over all conformations of the polymer chain. This necessitates the transformation of bond vectors into a common reference frame. In Figure 5.9 such a Cartesian reference frame is illustrated along a polymer backbone. To evaluate $l_{i-1} \cdot l_{i+1}$ in Figure 5.9, for example, l_{i+1} must be transformed into the reference frame along bond $i - 1$. This is accomplished with the matrix T_i which transforms a vector or a tensor from the reference frame along bond $i + 1$ to that along bond i (Eyring, 1932; Oka, 1947; Flory, 1969). As a consequence, l_{i+1} expressed along bond $i - 1$ is obtained from $T_{i-1} T_i l_{i+1}$.

$$T_i = \begin{vmatrix} \cos \theta_i & \sin \theta_i & 0 \\ \sin \theta_i \cos \phi_i & -\cos \theta_i \cos \phi_i & \sin \phi_i \\ \sin \theta_i \sin \phi_i & -\cos \theta_i \sin \phi_i & -\cos \phi \end{vmatrix}$$

Now $\langle r^2 \rangle_0$ can be obtained from $\langle r^2 \rangle_0 = 2Z^{-1} K^* \left(\prod_{i-1}^{n} G_i \right) K$, where

$$G_i = \begin{vmatrix} U_i & (U_i \otimes l_i^T)\|T\| & (l_i^2/2) U_i \\ O & (U_i \otimes E)\|T\| & U_i \otimes l_i \\ O & O & U_i \end{vmatrix}$$

and $K^* = [J^* \, 0 \, 0 \, \ldots \, 0](1 \times 5v)$, where v is the number of bond rotational states,

$$K = \begin{vmatrix} 0 \\ 0 \\ '' \\ '' \\ 0 \\ J \end{vmatrix} (5 \times 1)$$

l_i and l_i^T are the bond vectors in column and row representation, $l_i^2 = |l_i|^2$, E is the v order identity matrix, $\|T\|$ is the diagonal representation of T_i,

$$\|T_i\| = \begin{vmatrix} T(\phi_i = \alpha) & & & \\ & T(\phi_i = \beta) & & \\ & & \ddots & \\ & & & T(\phi_i = v) \end{vmatrix}$$

and $\otimes$ denotes the direct product of two matrices (Flory, 1969).

Any conformation-dependent property of a polymer chain can be calculated as a function of chain length by using the matrix methods outlined above if the following information is available for the polymer in question: chain geometry (bond lengths and angles), the RIS model (positions and energies of bond rotational states), and the contribution to the property made by each bond or attached group. In this manner, conformation-dependent properties, such as the mean-square end-to-end distance and dipole moment, optical anisotropy (as manifested by depolarization of scattered light, strain and electrical birefringence), optical rotatory power, equilibrium between cyclic and linear chains and between stereoisomers, conformational energies and entropies, and bond conformational properties useful in the analyses of ^{13}C NMR spectra, have been calculated and correlated with the appropriate experimental values [see next two chapters and Tonelli (1986)]. The temperature dependence of each property is also amenable to treatment and has been used to check the RIS models of several polymers (Flory, 1969; Mattice and Suter, 1994).

These matrix methods can easily accommodate both homo- and copolymers of any specified chemical and stereochemical composition and sequence distribution. Calculation of conformation-dependent properties with these matrix methods is readily accomplished with digital computers.

ENVIRONMENTAL EFFECTS ON POLYMER CONFORMATIONS

We have thus far treated the conformational characteristics of polymers by considering only short-range intramolecular interactions resulting in nearest-neighbor-dependent backbone bond rotational states. For polymers whose range of bond rotation interdependence goes beyond first neighbors, the statistical weight matrix U can be revised to connect the conformations of two (or more) bonds to the conformations chosen for the preceding pair (or triad, etc.) (Mattice and Suter, 1994). Such extensions are necessary and have been made for certain disubstituted vinyl polymers (Flory, Sundararajan, and DeBolt, 1974; Mattice and Suter, 1994). The possibility that segments remote along the sequence of a polymer chain might approach each other closely in space to interact and even overlap is ignored in the short-range RIS treatments of polymer conformations. These unaccounted-for long-range interactions—that is, involving pairs of units widely separated in the chain sequence—tend to expand the average polymer chain conformation in dilute solution, since compact conformations must lead to a greater number of long-range overlaps than occur in extended conformations. The expansion of polymer chain conformations in dilute solution by long-range interactions has been termed the *excluded volume effect* (Flory, 1953, 1969).

Excluded-volume perturbations of the chain conformation in solution depend on the effective covolume of a pair of chain segments in the given solvent. Thus, the expansion of a polymer chain depends on the solvent and the temperature. The poorer the solvent, clearly the smaller the chain expansion. By selecting the appropriate solvent and temperature, the segment volume can be compensated for by the mutual and favorable attractions between chain segments dissolved in the poor solvent. The excluded-volume effect then vanishes, because the effective covolume of the polymer segments goes to zero as the polymer segments seek out each other in preference to the poor solvent. Increasing the concentration of the polymer in solution to the point where their randomly coiling domains or spheres of influence or pervaded volumes overlap extensively should lead to a vanishing expansion effect from excluded volume. In bulk, amorphous polymers, the excluded volume effect is absent, because expansion of a polymer chain to avoid self-intersections only leads to increased numbers of equivalent intersections or interactions with other polymer chains, which are offsetting. Experimental evidence shows that the dimensions ($\langle r^2 \rangle_0$) of a polymer chain in the amorphous bulk and in dilute solution at the θ-point are generally closely similar (Wignall, 1996).

The temperature at which the polymer chain is no longer expanded owing to excluded-volume effects is called the theta-point (θ-point) or θ-temperature (Flory, 1953, 1969). An exact analogy to the θ-point in dilute polymer solutions can be found in the Boyle-point for real gases, where the repulsions and attractions between a pair of gas molecules compensate exactly to produce a zero covolume and ideal gas behavior. It is experimentally possible to conduct measurements at the θ-temperature, thereby eliminating the need to consider long-range perturbations of polymer conformations. Comparison of conformation-dependent properties measured in dilute solution at the θ-point can then be legitimately compared with those calculated with an RIS model based exclusively on short-range interactions, as described in the next chapter.

DISCUSSION QUESTIONS

1. Describe those features of polymers which make it possible to treat their conformational properties with the RIS model and the mathematical matrix methods introduced here.

2. How would you expect the response of a polymer to physical stimuli to be altered if the barriers to rotation between the rotational states of its backbone bonds were increased significantly from the ~ 3 kcal/mol barrier appropriate to alkanes like ethane and butane (see Figure 5.2)?

3. Suggest how our knowledge that the amount of volume V_i infuenced by a randomly coiling polymer sampling its myriad conformations is some two orders of magnitude larger than the volume V_0 actually physically occupied by its atoms and groups can explain the observation that very small amounts of high-molecular-weight polymers produce large increases in the viscosities of their solutions relative to the viscosity of the solvent used to make their solutions?

4. An elastic rubber band is composed of an amorphous collection of polymer chains that have been cross-linked into a three-dimensional network. The polymer sample would flow or be fluid without the cross-links. What inherent polymer property permits the rubber band to be easily stretched to several times its at rest

length and then snap back to its original length upon removal of the stretching force?

*5. Poly(isobutylene) (PIB) is a symmetric polymer like PE except that the hydrogens attached to alternate backbone carbons have been replaced by methyl groups. Derive an RIS model for PIB in terms of the statistical weights appropriate for the one-bond(σ) and two-bond (ω) interactions in PE. The U matrices for both $CH_2–C(CH_3)_2$ and $C(CH_3)_2–CH_2$ bonds in PIB must be developed. Newman diagrams about each bond will reveal the one-bond interactions (σ), and 5-carbon fragments centered on CH_2 and $C(CH_3)_2$, respectively, will reveal those two-bond pairs of adjacent backbone rotations that produce pentane interferences and incur statistical weights ω.

*6. Using the statistical weight matrices derived for PIB, compare the conformational flexibility of PIB to that of PE by comparing the U's of PIB to the U of PE. A comparison of the elements between the statistical weight matrices, after normalizing the largest elements in the PIB U's to 1, should make qualitatively apparent which is the conformationally more flexible polymer.

*7. A more quantitative comparison can be obtained by calculating the partition function Z for a two-bond fragment using the matrix multiplication method for each polymer. Remember Z provides a measure of the number of conformations available to a polymer, with each conformation weighted according to its energy by a Boltzmann factor.

*8. Compare the probability of finding each of the two different bonds in PIB in the t, g^+, g^- conformations with the 60% t, 20% g^+, 20% g^- bond conformer populations found previously for PE.

REFERENCES

Abe, A., Jernigan, R. L, and Flory, P. J. (1966), *J. Am. Chem. Soc.*, **88**, 631.

Bartell, L. S., and Kohl, D. A. (1963), *J. Chem. Phys.*, **39**, 3097.

Birshtein, T. M., and Ptitsyn, O. B. (1964), *Conformations of Macromolecules*, translated by Timasheff, S. N., and Timasheff, N. J. from the 1964 Russian Ed., Wiley-Interscience, New York.

Bonham, R. A., and Bartell, L. S. (1959), *J. Am. Chem. Soc.*, **81**, 3491.

Chandrasekhar, S. (1943), *Rev. Mod. Phys.*, **15**, 3.

Eyring, H. (1932), *Phys. Rev.*, **39**, 746.

Flory, P. J. (1953), *Principles of Polymer Chemistry*, Cornell University Press, Ithaca, New York.

Flory, P. J. (1969), *Statistical Mechanics of Chain Molecules*, Wiley-Interscience, New York.

Flory, P. J., Sundararajan, P. R., and De Bolt, L. C. (1974), *J. Am. Chem. Soc.*, **96**, 5015.

Herschback, D. R. (1963), *International Symposium on Molecular Structure and Spectroscopy, Tokyo, 1962*, Butterworths, London.

Hill, T. L. (1960), Introduction to Statistical Thermodynamics, Addison-Wesley, Reading, MA, Chap. 1.

Ising, E. (1925), *Z. Phys.*, **31**, 253.

Kramers, H. A., and Wannier, G. H. (1941), *Phys. Rev.*, **60**, 252.

Kuchitsu, K. (1959), *J. Chem. Soc. Jpn.*, **32**, 748.

Kuhn, W. (1934), *Kolloid Z.*, **68**, 2.

Mansfield, M. L. (1983), *Macromolecules*, **16**, 1863.

Mattice, W. L., and Suter, U. W. (1994), *Conformational Theory of Large Molecules*, Wiley-Interscience, New York.

Mizushima, S. (1954), *Structure of Molecules and Internal Rotation*, Academic Press, New York.

Newell, G. F., and Montroll, E. W. (1953), *Rev. Mod. Phys.*, **25**, 353.

Oka, S. (1942), *Proc. Phys. Math. Soc. Jpn*, **24**, 657.

Rayleigh, L. (1919), *Philos. Mag.*, **37**, 321.

Tonelli, A. E. (1985), *Macromolecules*, **18**, 2579.

Tonelli, A. E. (1986), *Encyclopedia of Polymer Science and Engineering*, N. Bikales, Ed., John Wiley and Sons, New York, p. 120.

Tonelli, A. E. (1989), *Nuclear Magnetic Resonance and Polymer Microstructure: The Conformational Connection*, VCH, New York.

Volkenstein, M. V. (1963), *Configurational Statistics of Polymer Chains*, translated from the Russian by Timasheff, S. N., and Timasheff, N. J., Wiley-Interscience, New York.

Wall, F. T. (1943), *J. Chem. Phys.*, **11**, 67.

Wignall, G. (1996), *Physical Properties of Polymers Handbook*, Mark, J. E., Ed., American Institute of Physics, Woodbury, New York, Chapter 22.

Wilson, E. B., Jr., (1959), *Adv. Chem. Phys.*, **2**, 367.

Wilson, E. B., Jr., (1962), *Pure Appl. Chem.*, **4**, 1.

CHAPTER 6

SOLUTION PROPERTIES OF POLYMERS

POLYMER EXCLUDED VOLUME

When a polymer sample is dissolved to form a solution, its molecularily dispersed chains are able to assume a vast array of conformations. As they randomly coil, they interact with solvent molecules and polymer segments, both their own and those belonging to other polymer chains. We generally expect, as illustrated in Figure 6.1, that in solution the dissolved polymer chains will be somewhat expanded beyond the $\langle \mathbf{r}^2 \rangle_0$ expected from its RIS conformational model. This is because some of the nearest-neighbor-dependent RIS conformations result in the close, through-space approach of polymer segments that are widely separated in sequence along the polymer backbone—say the 3rd and 68th segments, for example. From a physical point of view, not more than a single polymer segment can occupy the same volume. This implies that the volume occupied by a polymer segment is excluded from occupation by other polymer segments. Those RIS conformations (which only account for short-range intrachain interactions) that lead to excluded volume self-intersections, as in part (a) of Figure 6.1, will be precluded. Only expanded conformations, such as the one depicted in (b), are permited.

Consider a polymer dissolved in a poor solvent—that is, a solvent

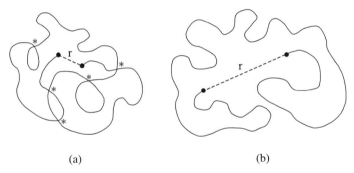

(a)   (b)

Figure 6.1. Isolated randomly coiling polymer chains with end-to-end distances, r. (a) Long-range intrachain excluded volume interactions shown at $*$; (b) no excluded volume interactions because of chain expansion.

where interactions between polymer segments are preferred over those between polymer segments and solvent molecules. Here the expansion of the polymer coil from excluded volume self-intersections may be compensated for by the desire of the polymer to adopt more compact conformations to reduce the unfavorable interactions between its segments and the solvent molecules. Now, only the short-range intramolecular interactions along the polymer chain, as accounted for in the appropriate RIS model, influence the overall macromolecular size and shape.

In a given solvent, the temperature at which excluded-volume expansion is exactly balanced by the solvent-induced contraction of the polymer chain is called the Flory or θ-temperature (Flory, 1953). This behavior is analogous to the Boyle temperature, T_B, of a real gas, where the gas behaves ideally according to $PV = nRT_B$. In dilute solution at the θ-temperature, the polymer chain is unperturbed by excluded-volume effects, and its average size ($\langle \mathbf{r}^2 \rangle_0$) and shape depend only on short-range intrachain interactions as described by its RIS model. As the concentration of polymer in solution is increased, the volumes influenced by each chain, V_i, begin to overlap and segments belonging to different chains begin to interact. As the overlapping of dissolved polymer chains increases, their expansion by excluded-volume self-intersections decreases, because expansion of each polymer coil to relieve self-intersections can only lead to a greater number of identical interactions between the segments belonging to different polymer coils. Thus, in more concentrated solutions, polymer chain conformations are unperturbed and not influenced by the solvent quality, and thus the polymer chains behave conformationally as in a dilute solution at the θ-temperature (Flory, 1953).

In very dilute solutions made with poor solvents, where polymer coils are well separated and free from overlap, as the temperature is lowered below the θ-temperature the polymer coils may contract below their unperturbed size to form polymer globules (Flory, 1953; Lifshitz, 1968; Grosberg and Kuznetsov, 1992). In this globular state, $\langle \mathbf{r}^2 \rangle \propto M^{2/3}$ rather than M for the unperturbed randomly coiling polymer, where $C_r = \langle \mathbf{r}^2 \rangle_0 / nl^2$, and so $\langle \mathbf{r}^2 \rangle_0 \propto (n$ or $M)$, or, as shown by Flory for the expanded chain at $T > \theta$, $\langle \mathbf{r}^2 \rangle \propto M^{6/5}$. As might be expected, further lowering of the temperature below θ generally produces aggregation of the polymer globules leading to precipitation. By controlling the solvent quality and temperature we can see that it is possible to alter the size of a polymer dissolved in a very dilute solution. As we will note in Chapter 8, this ability of a dissolved polymer chain to adjust its size has important consequences for those polymers necessary to sustain life—that is, biopolymers such as proteins and the poly(nucleic acids) DNA and RNA.

VOLUME INFLUENCED BY A RANDOMLY COILING POLYMER

The volume influenced or pervaded by a dissolved polymer chain, V_i, can be approximated by a sphere with a diameter $= (\langle \mathbf{r}^2 \rangle)^{1/2}$, or $V_i = (4\pi/3)(\langle \mathbf{r}^2 \rangle^{3/2})/8$. Of course in dilute solutions $\langle \mathbf{r}^2 \rangle$ will exceed $\langle \mathbf{r}^2 \rangle_0$ unless $T = \theta$ or beyond a certain polymer concentration, where coil overlap eliminates excluded-volume expansion of the polymer. Even with the lower limit of $\langle \mathbf{r}^2 \rangle = \langle \mathbf{r}^2 \rangle_0$, V_i is much larger than the hard-core volume, V_0, physically occupied by the constituent atoms of the polymer chain. V_0 can be estimated from the molecular weight and bulk density of the polymer. For example, a 10,000-bond PE chain has a RIS-calculated $(\langle \mathbf{r}^2 \rangle_0)^{1/2} = 400\,\text{Å}$ (Flory, 1969), or $V_i = 3.3 \times 10^7\,\text{Å}^3$, and a molecular weight of 140,000, which when considered with a bulk density of $\sim 1\,\text{g/cm}^3$, leads to a hard-core volume $V_0 = 2.3 \times 10^5\,\text{Å}^3$. Note that the ratio V_i/V_0, which we term the intimacy quotient IQ, exceeds 100. Consequently, the volume influenced

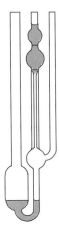

Figure 6.2. Schematic drawing of a Cannon–Ubbelhode viscometer.

by a high molecular weight, randomly coiling polymer is many times greater than the volume it physically occupies, or IQ $\gg$ 1. This fact explains why very small quantities of polymers produce large increases in the viscosities of their solutions compared with the viscosity of the pure solvent.

This can be readily illustrated with the viscometer shown in Figure 6.2. Such a viscometer is often used to characterize the dilute solution viscosity of polymer solutions. At room temperature the time taken for water to flow form the upper bulb through the capillary into the lower bulb is $\sim$2 min. (120 s) for a #75 viscometer, where the #75 specifies the capillary diameter. A 0.5 wt% aqueous ethanol (CH_3CH_2OH) solution shows a very similar flow time. By contrast, when 0.5 wt% poly(ethylene oxide) [PEO = $(-CH_2-CH_2-O-)_n$] with a molecular weight of 100,000 is dissolved in water the resulting solution exhibits a flow time of $\sim$3 min or 180 s. A measure of the increase in the viscosity of a solution over that of the pure solvent is $[t(\text{soln}) - t(\text{solv})]/t(\text{solv}) = \eta_{sp}$, where $t(\text{soln})$ and $t(\text{solv})$ are the solution and solvent flow times and η_{sp} is the specific viscosity. Note that $\eta_{sp} = 0.0$ for the ethanol solution and $\eta_{sp} = 0.5$ for the PEO solution. While a small amount (0.5 wt%) of ethanol does not affect the viscosity of water, the same small amount of PEO increases the flow time of water by $\sim$50%. Because of the similarity between the chemical structures of ethanol and the repeat unit of PEO, the great disparity in their abilities to effect the viscosities of their aqueous solutions must be a consequence of the great disparity in their molecular sizes—that is, ethanol with a molecular weight of 46 and PEO with a molecular weight of 100,000. This observation is all the more amazing when we remember that the ratio of ethanol:PEO molecules in the two aqueous solutions is $\sim 2000:1$.

INTRINSIC VISCOSITY OF A POLYMER SOLUTION

Let us consider a dissolved, randomly coiling polymer chain moving about in a solution dilute enough to prevent its overlap with other polymer chains. As the polymer coil moves, it must retard the motion of those solvent molecules that lie within the polymer's sphere of influence V_i. An illustration of this situation is presented in Figure 6.3, where the sizes of the arrows indicate the differences between the solvent velocities and the velocity of the polymer coil. This leads to the concept of an equivalent hydrodynamic sphere, impenetrable to solvent, which moves as a rigid bead or sphere of volume V_i (Flory, 1953). Einstein (1906, 1911) has shown that the specific viscosity η_{sp} as defined by $\eta_{sp} = (\eta_s - \eta_o)/\eta_o$, where η_s and η_o are the solution and solvent viscosities or flow times, is given by $\eta_{sp} = 2.5(n_2/V)U_e$, where n_2/V is the number of impenetrable beads or polymer coils per unit

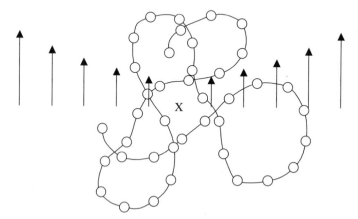

Figure 6.3. Translation (e.g., sedimentation) of a chain molecule with perturbation of the solvent flow relative to the molecule.

volume each with a volume U_e, and $U_e = V_i = (4\pi/3)(\langle \mathbf{r}^2 \rangle/2)^{3/2}$. Because $n_2/V = cN_A/100M$, where c is the concentration in grams per 100 ml, M is the molecular weight of the polymer coil, and N_A is Avogadro's number, $\eta_{sp}/c = [\eta] = 0.025N_A V_i/M$. Insertion of the explicit expression for V_i leads to the following expression for the intrinsic viscosity: $[\eta] = \{0.025N_A(4\pi/3)(\langle \mathbf{r}^2 \rangle/2)^{3/2}\}/M$.

Combining all constants in the above expression into a single constant Φ, leads to $[\eta] = \Phi\{(\langle \mathbf{r}^2 \rangle)^{3/2}\}/M$. If we assume that our polymer solution is at the θ-temperature, then the dissolved polymer coils are unperturbed by excluded-volume self-intersections and $\langle \mathbf{r}^2 \rangle = \langle \mathbf{r}^2 \rangle_0$, or $\langle \mathbf{r}^2 \rangle = \langle \mathbf{r}^2 \rangle_0 = C_r(nl^2)$, where the characteristic ratio of dimensions is given by $C_r = \langle \mathbf{r}^2 \rangle_0/nl^2$. Clearly for the unperturbed polymer chain $\langle \mathbf{r}^2 \rangle_0$ is proportional to the number of backbone bonds n or its molecular weight M. Substituting this relationship back into our expression for the intrinsic viscosity yields $[\eta] \propto M^{3/2}/M = M^{1/2}$. Thus we expect the intrinsic viscosity of a polymer solution to scale with the square root of its molecular weight when measured at $T = \theta$. This behavior has been observed on numerous occasions for a wide variety of different polymers (Flory, 1953; Brandrup et al., 1999).

Thus from measurement of the intrinsic viscosity of a polymer in a θ-solvent, the molecular weight of the sample can be obtained. Because intrinsic viscosiy measurements are relatively simple (see Figure 6.2), they are often employed to determine the molecular weights of polymer samples. Of course if the molecular weight of our polymer is already known, the dimensions of that same polymer in a θ-solvent can be obtained from the intrinsic viscosity, because $[\eta]_\theta = \Phi(\langle \mathbf{r}^2 \rangle_0/M)^{3/2}$. Of course if our polymer sample is dissolved in a good solvent, as opposed to a θ-solvent, then $\langle \mathbf{r}^2 \rangle \neq \langle \mathbf{r}^2 \rangle_0$, but rather $\langle \mathbf{r}^2 \rangle = \alpha^2 \langle \mathbf{r}^2 \rangle_0$. Here $\alpha > 1$ is the expansion factor of the unperturbed end-to-end distance caused by excluded-volume self-intersections. Under this circumstance $[\eta] = \Phi(\langle \mathbf{r}^2 \rangle_0/M)^{3/2}(M^{1/2})\alpha^3$, or $[\eta] = KM^{1/2}\alpha^3$, where $K = \Phi(\langle \mathbf{r}^2 \rangle_0/M)^{3/2}$ and should be independent of polymer molecular weight and the solvent. Further it can be demonstrated (Flory, 1953) that the expansion factor increases with the molecular weight of the polymer according to $\alpha^3 = M^{a'}$, so $[\eta] = KM^a$, where $a = 1/2 + a'$. The better the solvent, the greater the chain expansion and the larger is a'. At $T = \theta$, $\alpha = 1$, $a' = 0.0$, and $a = 0.5$. On the other hand, an upper limit (Flory, 1953) for a' is 0.3, so a would be $1/2 + 0.3$ or 0.8.

It is generally observed that in the same solvent at a single temperature $[\eta] = KM^a$, where $0.5 < a < 0.8$ for a wide range of molecular weights. Thus this relation can be used to determine the molecular weight of an unknown polymer sample provided that K and a have been previously determined for the same polymer in the same solvent at the same temperature on samples whose molecular weights were independently determined. Values of K and a have been determined for a wide range of polymers sometimes in a variety of solvents and for different temperatures (Brandrup et al., 1999). And so with a simple single intrinsic viscosity determination it is possible to establish the molecular weight of a polymer, without concern for finding a θ-solvent.

TURBULENT FLOW OF POLYMER SOLUTIONS

We have seen that the addition of small quantities of polymers to form dilute solutions produces a remarkable increase in their viscosities η_{sp} and $[\eta]$ compared with the pure solvents. This observation is universally true when observing the slow or "gentle" flow of polymer solutions—that is, such as in the case of the viscometer of Figure 6.2, where gravity produces the slow flow or passage of solution through a small capillary. However, when the diameter of the capillary is gradually inreased to allow the solution to flow faster, eventually the differences between the flow times of the solvent and the dilute polymer solutions begin to decrease, because the polymer coils become deformed and can no longer behave as spheres impenetrable to solvent (Fox et al., 1951).

The shear rate dependence of the intrinsic viscosity can be simply demonstrated by measuring the flow times of several dilute solutions with different concentrations of the same polymer through two viscometers with widely different capillary diameters. Selection of the viscometers should be based on the efflux time of solvent through each, with the narrow-bore viscometer leading to a solvent flow time in excess of several minutes, while the solvent flow time through the wide-bore viscometer should be a few tens of seconds. When the solution flow times are analyzed by extrapolation to infinite dilution, or as $c \to 0$, the intrinsic viscosities obtained for the same polymer solutions, but measured with these two viscometers, should differ. The intrinsic viscosity obtained with the wide-bore viscometer should be smaller, as if the polymers in the dilute solutions are of lower molecular weight and/or have smaller dimensions ($\langle \mathbf{r}^2 \rangle$). Of course neither the molecular weight nor the dimensions of the polymer are different in the two viscometers. Instead, in the shearing, solvent flow field of the wide-bore viscometer, the once-solvent impenetrable polymer coils become deformed and offer less resistance to the over all flow of the solution.

When taken to the extreme of highly turbulent or "rough" flow, dilute polymer solutions can even have viscosities lower than the pure solvents. This phenomenon, which is termed drag reduction, is not very well understood from a molecular point of view, but is nevertheless taken advantage of in several applications (Sellin and Moses, 1984; Kulicke et al., 1989; Hoyt, 1990). For example, if in the turbulent flow of water through a high-pressure fire hose, a small amount (in the ppm or 0.01% range) of a water-soluble polymer like the 100,000-molecular-weight PEO described above is added, the pressure required to obtain a given flow rate is dramatically reduced in comparison with pure water. In a related application, very dilute aqueous polymer solutions pumped just ahead of the bows of fast moving boats or ships can produce increases in their speeds or decreases in their fuel consumptions when a constant speed is maintained.

As we increase the concentrations of polymer solutions beyond where their random-coil spheres of influence begin to overlap, their viscosities (absolute viscosities in g/(cm-sec) = poise, not relative viscosities like η_{sp} and $[\eta]$) are seen to scale with $M\Phi_2$, where Φ_2 is the volume fraction of polymer in the solution. This behavior is even observed for concentrated polymer solutions up to and including bulk, liquid polymers ($\Phi_2 = 1$) (Berry and Fox, 1968), provided that their molecular weights are not too high (see subsequent discussion) and the flow is gentle and slow. Gentle and slow flows can be qualitatively distinguished from rough and fast flows by whether the flow takes place under a condition of zero shear rate or not, respectively. A liquid experiences shear when different portions of the liquid are moving at different velocities v, and γ the shear rate is simply the velocity gradient dv/dy in a particular direction y, or $\gamma = dv/dy$.

During our discussion of polymer solutions, their viscosities, and their flow behavior, we have not mentioned the specifics of forming polymer solutions. Because the process of dissolution and the resulting polymer solutions differ substantially from solutions made with small-molecule solutes, we now turn our attention to this subject.

POLYMER SOLUBILITY

Polymers are generally difficult to dissolve. They take longer to dissolve and only smaller amounts may be dissolved compared with low-molecular-weight solutes. Solubility is governed by two opposing factors: (1) the heat of mixing, which is a measure of the energy of interaction between solute and solvent molecules compared with solute–solute and solvent–solvent interaction energies, and (2) the entropy of mixing, which is a measure of the number of distinguishable arrangements of solute and solvent molecules in solution. Solubility increases when the heat of mixing is exothermic (heat given off as the energy content of the solution is reduced compared with pure solute and solvent) and the entropy of mixing is large. Those polymers whose repeat units have structures very similar to small molecules should also exhibit very similar heats of mixing. However, it is the loss in the entropy of mixing large, polymer solutes with small molecule solvents, as compared to the entropy of mixing obtained with small-molecule solutes, that lowers their solubilities (Flory, 1953; Morawetz, 1975).

This loss in mixing entropy can be illustrated with the help of Figure 6.4, where a 4×4 lattice is shown in (a) with eight solvent

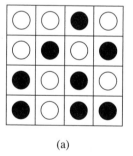   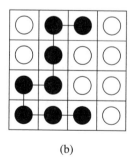

(a)   (b)

Figure 6.4. The mixing of eight solvents (○) and eight solute (●) molecules on a 4×4 lattice (a), and an eight-segment polymer (–●–●–●–) and eight solvent (○) molecules also on a 4×4 lattice (b).

(○) and eight solute (●) molecules placed in one of several distinguishable arrangements. It can be shown that in all there are $16!/(8!8!) = 12,870$ distinguishable ways of arranging the eight solvent and eight solute molecules on the 16-site, 4×4 lattice. (Remember that the number of combinations of n_0 things taken n_1 at a time is $n_0!/[n_1!(n_0 - n_1)!]$.) However, once the eight solute molecules are bonded together to form an eight-membered chain, the number of distinguishable arrangements with the eight solvent molecules is reduced by more than a factor of 40 to 300. Connecting the eight solute molecules into a chain causes them to lose their indistinguishibilty; that is, they are now distinguishable (DiMarzio, 1999), thereby greatly reducing the number of ways they maybe arranged with solvent and thus their entropy of mixing. This loss of mixing entropy, or really the failure to gain entropy on mixing, requires a much more favorable energy of interaction between polymer segments and solvent molecules to achieve polymer dissolution than is the case for small-molecule solutes. Consequently, polymers usually exhibit solubilities that are much reduced from those of small molecules with chemical structures very similar to those of the polymer repeat units. [Some rainy Saturday afternoon, when your cable is out, your computer is down, your car will not start, and you have reached a temporary point of saturation reading about polymers in this book, we suggest that you take a pad of paper and a pencil and convince yourself that there are "only 300" distinguishible ways to arrange eight solvent molecules and a single eight-membered polymer chain on a $4 \times 4 = 16$ site square lattice.]

ANISOTROPIC, LIQUID CRYSTALLINE POLYMER SOLUTIONS

When certain polymers are dissolved to form solutions, it is observed that their solutions no longer remain isotropic above a certain polymer concentration. Instead they form a liquid crystalline mesophase commonly referred to as a lyotropic liquid crystal (LC). The tobacco mosaic virus (TMV) was one of the first polymeric materials known to form a liquid crystal phase (Bawden, et al., 1936) In such solutions the polymer chains are no longer equally oriented in all directions, i.e., isotropically, but have long-range orientational order with polymer chain axes more or less oriented parallel to each other in the LC solution. Conformationally stiff, rod-like polymers, often possessing mesogenic or planar and rigid fused ring structures in their backbones or attached as side chains, cannot be arranged isotropically in solution above a certain concentration, where the rod-like polymers or mesogenic groups begin to pack anisotropically and order within microscopic domains. (This behavior may be simulated by attempting to achieve an isotropic arrangement of pick-up sticks or uncooked spaghetti in progressively smaller containers, where they will eventually be observed to fit only through their tight, ordered, anisotropic packing.) Though the LC polymer chains are ordered within microscopic domains, these domains may remain disordered with respect to each other. Figure 6.5 presents a schematic illustration of LC polymers, and their modes of local ordering to produce anisotropic domains (Papkov, 1984; Ober et al., 1984; Srinivasarao, 1995). Whether the local ordering of polymer chains in lyotropic LC polymer solutions is produced by their inherent conformational rigidity (Onsager, 1949; Flory, 1954) or by the interaction between mesogenic groups in their backbones or side chains, they are not able to randomly coil, but instead adopt anisotropic, extended conformations. The flow behavior or rheology of LC polymer solutions are consequently distinct (and complex) from solutions containing randomly coiling polymer chains (Srinivasarao, 1995).

In the remaining discussion we touch upon means for measuring the molecular weights and sizes of polymers in solution and the time dependence of their flowing solutions.

POLYMER MOLECULAR WEIGHTS MEASURED IN SOLUTION

We have seen that the molecular weight of a polymer sample may be assessed from intrinsic viscosity measurements. Here brief mention is made of several other experimental means to determine the molecular weights of polymers. In principle, observation of the colligative properties of dilute polymer solutions should provide a means for determining the molecular weight of macromolecular solutes. The magnitudes of the freezing point depression, boiling point elevation, and osmotic pressure produced by the addition of a solute to a liquid to

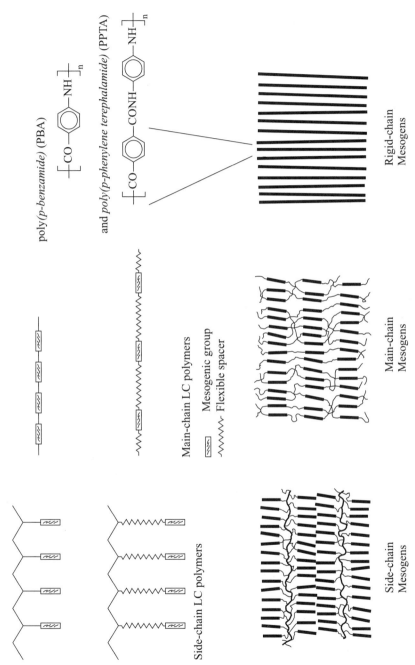

Figure 6.5. Schematic representation of macromolecular liquid crystals.

TABLE 6.1. Comparison of Calculated Boiling Point Elevation, Freezing Point Depression, and Osmotic Pressure (Flory, 1953)

M	$(\Delta T_b/c)_0^a$ in °C/(g/100 ml)	$(\Delta T_f/c)_0^a$ in °C/(g/100 ml)	$(\pi/c)_0$ in (g/cm^2)/(g/100 ml)
10,000	0.0031	0.0058	25
50,000	0.0006	0.0012	5
100,000	0.0003	0.0006	2.5

^{a}Values for benzene.

form a solution are proportional to the number of moles of added solute. Because the number of moles of solute added to form a solution is directly/inversely proportional to the weight/molecular weight of the solute, the magnitude of each colligative property is inversely proportional to the solute's molecular weight.

Osmotic Pressure of Polymer Solutions

However, as is apparent from Table 6.1, the magnitudes of the colligative properties are very small for freezing point depression and boiling point elevation, but are significantly larger for the osmotic pressure when the solute molecular weights are large. As a result, among the colligative properties of polymer solutions only the osmotic pressure can be used to estimate the molecular weights of macromolecular solutes with $\overline{M}_n > 10,000$. The osmotic pressure π exerted by the solvent surrounding the polymer solution contained inside the semipermeable membrane shown in Figure 6.6 is given by $(\pi/c)_0 = RT/\overline{M}_n$, where c is the concentration of the solution. π is manifested by the height of the liquid column above the polymer solution, which is created by solvent diffusing across the polymer impermeable membrane into the solution in order to lower its free energy. For a 0.001 molar solution, π corresponds to a 10-in. column of water produced by dissolving 1 g of polymer with a $\overline{M}_n = 10,000$ in 100 ml of solvent. (π/c) is measured for several dilute solutions and is extrapolated to $c \rightarrow 0$ to obtain $\overline{M}_n$.

Light Scattering from Polymer Solutions

When light is passed through a dilute polymer solution made with a solvent whose refractive index differs from that of the dissolved poly-

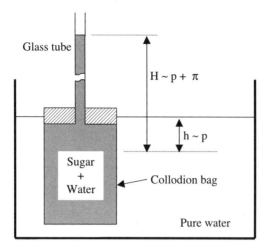

Figure 6.6. Simple osmotic pressure experiment.

mer, the light will be scattered with an intensity that depends on both the size and concentration of dissolved polymers in addition to the difference in refractive indices of the polymer and the solvent (Debye, 1944, 1947; Flory, 1953). If the solution is sufficiently dilute, the intensity of the scattered light will be a sum of scattering contributions made by each of the well-separated polymer coils. When the dissolved polymer coils are significantly smaller than the wavelength of the light, which is usually the case for all but the highest-molecular-weight polymers, and they are isotropic or have the same polarizability in all directions, then the intensity of light scattered in any direction by each polymer coil will be proportional to the square of its size. This is because the intensity of the scattered light is proportional to the square of the amplitude of the scattered light wave vector, which in turn is proportional to the square of the excess polarizability of the polymer over the solvent. The excess polarizability is proportional to $c \, (dn/dc)$, where c and n are the concentration and refractive index of the solution (Smoluchowski, 1908, 1912; Einstein, 1910) and dn/dc is the refractive index increment of the solution. [Without a refractive index contrast between the dissolved polymer and the solvent, contributions made to light scattering from the dissolved polymer chains vanish.] The scattered intensity is proportional to c^2 and M^2, because $c = NM/N_A V$, where N/V is the number of polymer coils per unit volume and N_A is Avogadro's number.

Because the intensity of light scattered by each polymer coil depends on the square of its size, and recalling from Chapter 2 that $\bar{M}_w = \sum N_x M_x^2 / \sum N_x M_x$, light scattering observed from dilute polymer solutions permits an evaluation of the weight-average molecular weight $\bar{M}_w$. In combination with osmotic pressure measurements, light scattering can assist in providing a measure of the molecular weight distribution of a polymer sample in the form of its polydispersity index $\text{PDI} = \bar{M}_w / \bar{M}_n$.

The intensity of light scattered from dilute polymer solutions at various angles from the incident light beam may also be utilized to determine the dimensions of the dissolved polymer coils $\langle \mathbf{r}^2 \rangle$ (Zimm, 1948; McIntyre and Gornick, 1961). If the light-scattering measurements are performed in a θ-solvent, then the dimensions obtained are unperturbed by excluded volume self-intersections or $\langle \mathbf{r}^2 \rangle = \langle \mathbf{r}^2 \rangle_0$ and may be compared to those calculated from the RIS conformational model appropriate to the polymer under study. Even if the polymer solution is not at $T = \theta$, the unperturbed polymer coil dimensions $\langle \mathbf{r}^2 \rangle_0$ can be estimated, because the expansion factor of the dimensions $\alpha = (\langle \mathbf{r}^2 \rangle / \langle \mathbf{r}^2 \rangle_0)^{1/2}$ can be derived from an extrapolation of the angular-dependent scattered light intensities to a scattering angle of $0°$ (Zimm, 1948; McIntyre and Gornick, 1961).

Clearly light-scattering observations made on dilute polymer solutions using solvents with refractive indices substantially different from the dissolved polymers serve as an important means for characterizing them. Not only can the weight-average molecular weight $\bar{M}_w$ of the sample be determined, but the physical dimensions of the dissolved polymer coils $\langle \mathbf{r}^2 \rangle$ and a measure of the solvent quality for the polymer, as provided by the expansion factor α, are also obtainable from the light-scattering measurements.

Intrinsic viscosity, osmotic pressure, and light-scattering measurements of dilute polymer solutions have been briefly introduced. We have noted that the viscosity-average $(\bar{M}_v)$, the number-average $(\bar{M}_n)$, and the weight-average $(\bar{M}_w)$ molecular weights of polymer samples, respectively, are yielded by these three polymer characterization methods. $\bar{M}_n = \sum N_i M_i / \sum N_i$, where N_i and M_i are the number and molecular weight of species i, and $\bar{M}_w = \sum N_i M_i^2 / \sum N_i M_i$. The intrinsic viscosity $[\eta] = K' M^a$, where a normally ranges from 0.5 to 0.8. A heterogeneous polymer in dilute solution will have a specific viscosity given by $\eta_{sp} = \sum (\eta_{sp})_i$, if each polymer contributes independently to the solution viscosity. $(\eta_{sp})_i = K'(M_i^a)c_i$, where c_i and M_i are the concentration and molecular weight of the same species. Then $\eta_{sp} = K'\sum(M_i^a)c_i$, or $[\eta] = \eta_{sp}/c = K'\sum(M_i^a)c_i/c$, where c is the total concentration of all polymer species.

The viscosity-average molecular weight may then be defined as $\bar{M}_v = [\sum w_i M_i^a]^{1/a} = [(\sum N_i M_i^{(1+a)}/\sum N_i M_i]^{1/\alpha}$, where the weight fraction of species i is $w_i = c_i/c$ and N_i is the number of species i molecules (Flory, 1953). Now we can compare $\bar{M}_n$, $\bar{M}_v$, and $\bar{M}_w$: $\bar{M}_n = \sum N_i M_i/\sum N_i$; $\bar{M}_v = [(\sum N_i Mi_i^{(1+a)})/\sum N_i M_i]^{1/a}$ and $\bar{M}_w = \sum(N_i M_i^2)/\sum N_i M_i$ Because $a = 0.5$–0.8, $\bar{M}_n < \bar{M}_v < \bar{M}_w$, where $\bar{M}_v$ is generally closer to $\bar{M}_w$ for most polymer samples. In the next chapter we will learn that certain physical properties of bulk polymers are more closely dependent on $\bar{M}_w$, or $\sim \bar{M}_v$, than on $\bar{M}_n$.

ESTIMATION OF POLYMER DIMENSIONS IN SOLUTION

Viscosity and light scattering measurements performed on dilute polymer solutions reflect the sizes $(M, \langle \mathbf{r}^2 \rangle)$ of the dissolved polymer chains and can be utilized to test the conformational characteristics (RIS models) of polymers, which of course depend on their detailed microstructures. However, most flexible, randomly-coiling polymers have dimensions $C_r = \langle \mathbf{r}^2 \rangle_0/(nl^2)$ within a factor of 2–3 of each other, so the conformationally averaged overall size of a polymer is not very sensitively dependent on its microstructure. There is a property, however, when averaged over all conformations accessible to a polymer chain, which is quite sensitive to the detailed polymer microstructure. That property is the molar Kerr constant, $_mK$, determined by observing the birefringence, in excess of the solvent, induced in a dilute polymer solution when an electric field is applied (Flory, 1969; Riande and Saiz, 1992). [The solvents must necessarily have nearly vanishing dipole moments and be nearly isotropically polarizable.]

Kerr first observed the birefringence (difference in refractive indices parallel and perpendicular to the applied electric field) produced upon application of an electric field in 1880 (Kerr, 1880). The electric field acts upon the dissolved polymer chains both through their overall permanent electric dipole moments and through the anisotropy of their overall polarizabilities, thereby aligning the polymer chains and producing an overall, anisotropic arrangement of the dissolved polymers, which is manifested by the solution birefringence. It can be shown (Nagai and Ishikawa, 1965; Flory, 1969) that the first contribution is given by $\langle \mu^R \hat{\alpha} \mu^C \rangle$ and the second by $\langle \hat{\alpha}^R \hat{\alpha}^C \rangle$, where μ and $\hat{\alpha}$ are the overall dipole moment vector and anisotropic polarizability tensor expressed in row (R) and column (C) forms, respectively. Just as the average dimensions of a dissolved polymer depend on the end-to-end distance of each available chain conformation appropriately averaged over the complete set of conformations accessible to the polymer chain (see Chapter 5), so too must the dipole moment and anisotropic polarizability contributions made to the molar Kerr constant $_mK$ be averaged over all available conformations.

Each of these terms may be conformationally averaged and calculated from its RIS conformational model by means of the matrix multiplication method, as illustrated previously, when evaluating Z and $\langle \mathbf{r}^2 \rangle_0$ for a polymer chain, except that the generator matrix G_i for each bond must not only contain U_i and $\|T_i\|$, but also the dipole moment vector m_i and the anisotropic polarizability tensor $\hat{\alpha}_i$ for each backbone bond i. Then the molar Kerr constant is given by $_mK = (2\pi N_A/135)[\langle \mu^R \hat{\alpha} \mu^C \rangle/(kT^2) + \langle \hat{\alpha}^R \hat{\alpha}^C \rangle/(kT)]$ (Flory, 1969). In Figure 6.7 the molar Kerr constants calculated for styrene-p-bromostyrene copolymers of varying compositions and stereosequences are displayed along with those measured for a series of atactic styrene-p-bromostyrene copolymers (Khanarian et al., 1982). These copolymers were obtained from the random bromination of an atactic polystyrene sample by addition of varying amounts of bromine to a nitrobenzene solution of the polystyrene and stirring for 24 hr in the dark. Under these bromination conditions, all bromines were observed to be attached exclusively at the para-positions on the phenyl ring.

Note the extraordinary sensitivity of the calculated $_mK/x$ to both composition and tacticity, which covers a range of three orders of magnitude and also changes sign. This is particularily true for high *para*-bromostyrene contents. Comparison with measured $_mK$'s indicate that the tacticty of the starting polystyrene and the resultant brominated copolymers is atactic with $P_r = 0.55 \pm 0.05$, or 55% randomly placed racemic (r) diads. Subsequent assignments of the ^{13}C NMR spectra of atactic polystyrenes, based on the ^{13}C NMR spectra of pentad and hexad model compounds of polystyrene, also concluded that free-radical polymerized polystyrenes are atactic with $\sim 55\%$ randomly placed racemic diads (Sato et al., 1982).

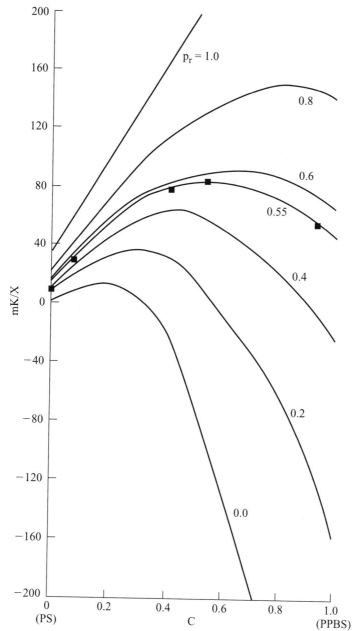

Figure 6.7. $_mK/x$ ($\times 10^{-12}$ cm^7 SC^{-2} mol^{-1}) of the copolymer as a function of composition c and tacticity p_r. Black squares are experimental results. (Khanarian et al., 1982)

Because bromination of the phenyl rings in the *para*-position is not expected to alter the conformational properties (RIS model) of polystyrene (Yoon et al., 1975), the styrene-*p*-bromostyrene copolymers produced by bromination of polystyrene differ only in the dipole moments and polarizabilities of their pendant phenyl side-chains. This example and others serve to recommend comparison of measured and calculated molar Kerr constants as a very sensitive means to test both the microstructural and conformational charateristics of polymers (Tonelli, 1976, 1986).

SIZE-EXCLUSION CHROMATOGRAPHY

Our discussion of the means to determine the molecular weights and sizes of polymers concludes with a method called gel permeation or size-exclusion chromatography often given the acronyms GPC and SEC, respectively. This method relies on the ability of a column packed with a porous gel—that is, a cross-linked polymer network swollen with solvent—to retard the flow of dissolved polymers to different degrees based on their sizes or molecular weights. Since it is generally observed that smaller, lower-molecular-weight polymers experience more retardation and take longer to pass through such a column than larger, higher-molecular-weight polymers, it has been assumed that more of the pores in the stationary swollen gel are accessible to and visited by the short chains, while the longer chains are too large to penetrate all but the largest of these pores and, therefore, pass more quickly through the packed column (Yau et al., 1979). Figure 6.8 presents a micrograph of a Styragel (10^6) porous packing medium on which two polystyrene coils of 100,00 and 1,000,000 molecular weight have been superimposed (Yau et al., 1979). It seems apparent that the lower molecular weight polystyrene coil can sample more pores and would experience a greater retardation of its elution through the column than the higher-molecular-weight polystyrene coil.

Figure 6.9 presents a schematic of the GPC/SEC experiment with the accompanying elution curve, which describes the time dependence of the concentration of polymer leaving the column. Various means, including differential refractometers and viscometers and low-angle light-scattering, UV, and IR detectors, have been employed to estab-

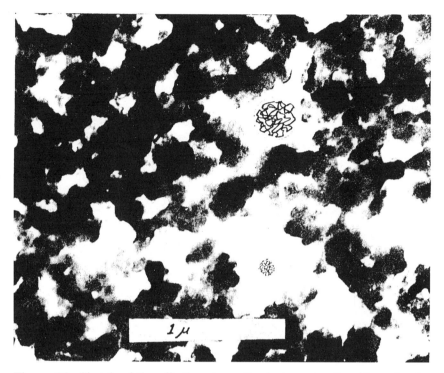

Figure 6.8. Sketch of the effective sizes of polystyrene coils with molecular weights 100,000 and 1,000,000 in THF on an electron photomicrograph of a Styragel (10^6) column packing.

lish the elution curves. To convert the elution curve (concentration of polymer in eluent versus time) to a molecular weight distribution curve (concentration or weight fraction versus molecular weight) the elution of standard polymers with independently determined, monodisperse molecular weights (PDI ~ 1) that cover a broad range of molecular weights must be observed for each GPC/SEC column. A set of monodisperse polystyrenes are generally used for column calibration. Once the column is calibrated, the molecular weight distribution (N_i or w_i versus M_i) of a polymer sample can be determined, from which the number- and weight-average molecular weights $\overline{M}_n$ and $\overline{M}_w$ can be calculated.

GPC/SEC is a convenient method for obtaining the distribution of molecular weights in a polymer sample, which viscosity, osmometry,

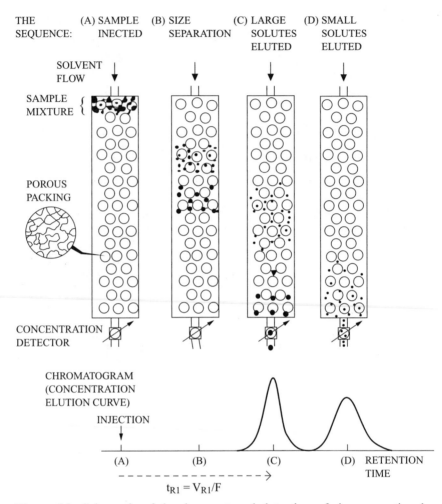

Figure 6.9. Schematic of development and detection of size separation by GPC/SEC.

and light-scattering measurements alone cannot provide, and is made possible by the fractionation or separation of the sample into its individual components according to size. However, the calibration of GPC/SEC columns with known polymer molecular weight standards renders the results relative and not absolute, unless the polymer sample under study consists of the same type of polymer used to calibrate the column. Nevertheless, GPC/SEC has generally supplanted all other methods to become the most commonly used method to study the molecular weights of polymer samples.

TIME-DEPENDENCE OF FLOWING POLYMER SOLUTIONS

Before closing the discussion in this chapter, we revisit the flow behavior of polymer solutions. Mention has already been made of the dramatic decrease in the flow rates of dilute polymer solutions below those exhibited by their pure solvents in capillary viscometers. On the other hand, when equally small or even reduced amounts of polymers are dissolved to form solutions that are subjected to very vigorous, rapid, turbulent flows, these polymer solutions may in fact exhibit flow times below (flow rates above) those of the pure solvents, a phenomenon termed *drag reduction*. This comparison of polymer solution fluidity, where under slow, gentle flow conditions the polymer's presence retards the flow and under fast, turbulent flow the polymer enhances or lubricates the flowing solution, points out the generally observed time-dependent response of polymer materials, both in solution and the bulk (see next chapter).

This characteristic feature of polymers is illustrated in Figure 6.10, where the viscosities of poly(vinyl acetate)/diethyl phthalate solutions of different concentrations (Berry and Fox, 1968) are displayed and plotted versus the product of their molecular weights M and their concentrations Φ_2 (volume fraction), ie., $M\Phi_2$, in (a), where $Z_w \alpha M$, and for different solution flow or shear rates for a single polystyrene/ toluene solution (Graessley, 1974) in (b). Note that in (a) for slow flows (zero shear rate) η scales and increases with $M\Phi_2$, while for the faster flows (high shear rates) in (b) the viscosity of the single polystyrene solution begins to decrease. These observations remind us of the contrasting time dependent flow behaviors of dilute polymer solutions seen in the slow capillary flows of intrinsic viscosity measurements and the fast turbulent flows encountered in drag reduction applications.

Note that in Figure 6.9(a) the zero shear rate viscosities change from a linear (1.0) to a 3.4 power dependence upon $\Phi_2 M$. Though discussed more fully in the next chapter, we mention here that this behavior is believed (Graessley, 1974; Lin and Juang, 1999) to be the result of entanglements between chains. Once entangled, the flow rate

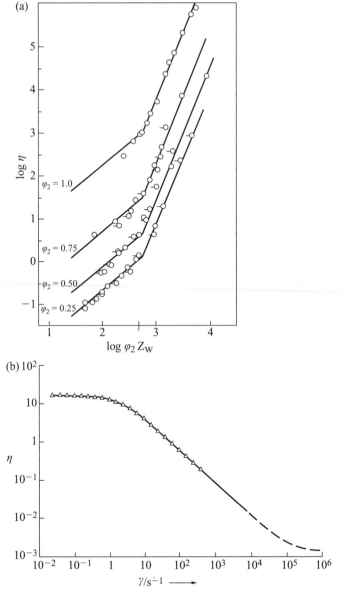

Figure 6.10. (a) The zero shear viscosities of poly(vinyl acetate)/diethyl phthalate solutions of varying concentrations and molecular weights (Berry and Fox, 1968). (b) The shear rate dependence of the viscosity of a polystyrene/toluene solution ($\overline{M}_w = 23.6 \times 10^6$, $c = 0.1$ g/ml) (Graessley, 1974).

of the solution, or melt, is no longer simply proportional to the frictional coefficient between polymer segments and solvents, or other polymer segments, but in addition the chains must at least partially unentangle during flow. This is why a higher molecular weight is required to evidence entanglement as the polymer concentration Φ_2 is decreased.

DISCUSSION QUESTIONS

1. Discuss the conformations and sizes of polymer chains in bulk, amorphous polymer samples, such as molten polymer liquids, in terms of the unperturbed values obtained from their RIS conformational models.

2. Why must the capillary flow times used to determine the intrinsic viscosity of a polymer in a particular solvent at a specified temperature be measured for a series of dilute polymer solutions with different concentrations and then be extrapolated to infinite dilution?

3. Given two samples of the same polymer (polyA1 and polyA2) and a sample of a chemically distinct polymer (polyB), what can be safely concluded concerning their molecular weights if polyA2 and polyB show the same intrinsic viscosities in the same solvent at the same temperature, which is larger than that measured for polyA1?

4. Would you expect any difference in the solubilities of two samples of the same polymer whose molecular weights are 10,000 and 1,000,000? Please give reasons for your answer.

5. Compare the IQ's for a polymer in very dilute solution at $T = \theta$, where it is an unperturbed random coil, and at $T < \theta$, where it has collapsed into a globule.

*6. Compare and contrast the determination of polymer molecular weights M and sizes $\langle r^2 \rangle$ by means of intrinsic viscosity, osmotic pressure, and light-scattering methods. Be sure to assess the relative advantages and disadvantages of each method in terms of both experimental difficulties and the types of information obtainable.

*7. After constructing a RIS model to describe the conformational characteristics of a polymer, it must be tested by calculating properties of the polymer, such as $\langle r^2 \rangle_0$, which depend on and must be averaged over all conformations accessible to the polymer. Why is the molar Kerr constant $_mK$ of a polar polymer a more appropriate conformationally sensitive property with which to test its RIS model than say $\langle \mathbf{r}^2 \rangle_0$?

*8. What is the major weakness in the GPC/SEC determination of the molecular weight distribution of a polymer sample? If possible, explain how this shortcoming can be remedied?

*9. Notice in Figure 6.9(a) that for each solution concentration Φ_2 the viscosity scales as M to the first power for $M < 14,000$ and as M to the 3.4 power for larger molecular weights. Also notice that these solutions are very concentrated (volume fraction of polymer $= \Phi_2 = 0.25$–1.0). Recall our definition of the intimacy quotient IQ of a polymer chain, which is given simply by the ratio of pervaded volume or volume influenced V_i by the randomly coiling polymer to that of the hard-core volume V_0 of its constituent atoms and speculate on the cause of the abrupt rise in the dependence of the viscosities of these solutions upon the molecular weights of their dissolved polymer chains.

REFERENCES

Berry, G. C., and Fox, T. G. (1968), *Adv. Polym. Sci.*, **5**, 261.

Brandrup, J., Immergut, E. H., and Grulke, E. A. (1999), *Polymer Handbook*, 4th ed., Wiley-Interscience, New York.

Debye, P. (1944), *J. Appl. Phys.*, **15**, 338.

Debye, P. (1947), *J. Phys. and Colloid Chem.*, **51**, 18.

DiMarzio, E. A. (1999), *Prog. Polym. Sci.*, **24**, 329.

Einstein, A. (1906), *Ann. Physik.*, **17**, 289.

Einstein, A. (1910), *Ann. Physik.*, **33**, 1275.

Einstein, A. (1911), *Ann. Physik.*, **34**, 591.

Ferry, J. D. (1970), *Viscoelastic Properties of Polymers*, 2nd ed., Wiley, New York.

Flory, P. J. (1953), *Principles of Polymer Chemistry*, Cornell University Press, Ithaca, New York.

Flory, P. J. (1954), *Proc. R. Soc. (London)*, **A234**, 73.

Flory, P. J. (1969), *Statistical Mechanics of Chain Molecules*, Wiley-Interscience, New York.

Fox, T. G., Jr., Fox, J. C., and Flory, P, J. (1951), *J. Am. Chem Soc.*, **73**, 1901.

Graessley, W. W. (1974), *Adv. Polym. Sci.*, **16**, 1.

Grosberg, A. Yu., and Kuznetov, D. V. (1992), *Macromolecules*, **25**, 1970.

Hoyt, J. W. (1990), *Concise Encyclopedia of Polymer Science and Engineering*, J. I. Kroschwitz, Ed., John Wiley and Sons, New York, p. 278.

Kerr, J. (1880), *Philos. Mag. Ser.*, **9**, 157.

Khanarian, G., Cais, R. E., Kometani, J. M., and Tonelli, A. E. (1982), *Macromolecules*, **15**, 866.

Kulicke, W.-M., Kotter, M., and Grager, H. (1989), *Adv. Polym. Sci.*, **89**, 1.

Lifshitz, I. M. (1968), *Zh. Eksp. Teor. Fiz.*, **55**, 2408 [English translation in *Soviet Physics. JETP*, **28** 1280 (1969)].

Lin, Y.-H., and Juang, J.-H. (1999), *Macromolecules*, **32**, 181.

McIntyre, D., and Gornick, F. (1961), *Light Scattering from Dilute Polymer Solutions*, Gordon and Breach, New York.

Morawetz, H. (1975), *Macromolecules in Solution*, 2nd ed., Wiley-Interscience, New York.

Nagai, K., and Ishikawa, T. (1965), *J. Chem. Phys.*, **43**, 4508.

Ober, C. K., Jin, J.-I., and Lenz, R. W. (1984), *Adv. Polym. Sci.*, **59**, 103.

Onsager, L. (1949), *Ann. NY Acad. Sci.*, **51**, 627.

Papkov, S. P. (1984), *Adv. Polym. Sci.*, **59**, 75.

Riande, E., and Saiz, E. (1992), *Dipole Moments and Birefringence of Polymers*, Prentice Hall, Englewood Cliffs, NJ.

Sato, H., Tanaka, Y., and Hatada, K. (1982), *Makromol. Rapid Commun.*, **3**, 175, 181.

Sellin, R. H. J., and Moses, R. T. (1984), *Proceedings of the 3rd International Conference on Drag Reduction, Bristol, UK*, University of Bristol, Bristol, UK.

Smoluchowski, M. (1908), *Ann. Physik.*, **25**, 205.

Smoluchowski, M. (1912), *Philos. Mag.*, **23**, 165.

Srinivasarao, M. (1995), *Int. J. Mod. Phys*, B9, 2515.

Tonelli, A. E. (1976), *Macromolecules*, **10**, 153.

Tonelli, A. E. (1978), *Chemistry*, **51**, 11.

Tonelli, A. E. (1986), *Am. Chem. Soc. Meeting Abstracts*, **73**, 192.

Yau, W. W., Kirkland, J. J., and Bly, D. D. (1979), *Modern Size-Exclusion Chromatography*, John Wiley and Sons, New York.

Yoon, D., Sundararajan, P. R., and Flory, P. J. (1975), *Macromolecules*, **8**, 776.

Zimm, B. H. (1948), *J. Chem Phys.*, **16**, 157.

CHAPTER 7

BULK PROPERTIES OF POLYMERS

INTRODUCTION

In the bulk, polymer chains are in close contact with other polymer chains. Two extreme situations can arise: (1) In amorphous samples the conformationally disordered, randomly coiling polymer chains are in contact with many other polymer chains, on the order of 100 of their neighbors based on a typical IQ $= V_i/V_0$ value, and (2) in crystalline samples the highly extended chains have an IQ approaching unity and are in intimate contact with a much smaller number of neighboring polymer chains (4–6). Although fewer in number, the interactions between neighboring polymer chains in a crystal are stronger, are more intimate, and extend over much longer portions of their chain contours in comparison to the more numerous, yet less extensive, contacts between the pervasive, coiling polymer chains of an amorphous sample. Because of their long-chain natures, polymers, when subject to stimuli, may most efficiently respond internally by altering their conformations. However, this intramolecular, conformational response may be retarded or even precluded in the bulk by the constraints due to the interactions between proximal chains. Consequently, though we can expect an internal conformational response from a polymer under stress, we cannot expect the speed or frequency nor the size or amplitude of the response to be governed solely by the intrachain constraints to internal backbone bond rotations encountered in an isolated polymer, say in dilute solution. Rather, in the bulk both the frequency and amplitude of the conformational responses of the constituent chains are generally dominated by cooperative interactions between them.

POLYMER LIQUIDS

We begin our discussion of bulk polymer properties by turning our consideration toward polymer liquids—that is, disordered samples of intertwined, randomly coiled polymers possessing sufficient thermal energy to evidence irreversible plastic flow if given sufficient time to respond to an applied stress, such as the force of gravity, for example. When a polymer liquid flows the center of mass of each coiled chain must translate, and the relative displacements between chains are altered. How can this occur when each coiled chain in a polymer liquid is in close contact with many neighboring chains due to its large pervaded volume V_i and $IQ = V_i/V_0$? The answer clearly is not by translations of conformationally fixed or rigid polymer chains, because the multiple contacts or entanglements between random coils would require the entire sample to translate with no alteration of the relative separations between chains. For a bulk polymer sample to flow, each of its constituent chains must be able to change conformations so as to permit adjustments of the interactions or entanglements between chains that accompany their relative displacements. In other words, because of extensive overlap between the volumes influenced by each chain, which does not occur in small molecule liquids where $IQ \sim 1$, the flow of polymer liquids is a much more cooperative process.

In Figure 7.1 we see the viscosities of several liquid polymers measured in slow flows (zero shear rate) plotted as a function of their molecular weights (Graessley, 1974). We note that their zero shear viscosities scale with the first power of their molecular weights up to a critical molecular weight M_c, and above M_c there is an abrupt change in slope reflecting a dependence of viscosity on the 3.4 power of their molecular weights. Apparently above a certain chain length or molecular weight the overlap and entanglement of neighboring chains is sufficient to produce an abrupt increase in the dependence of the vis-

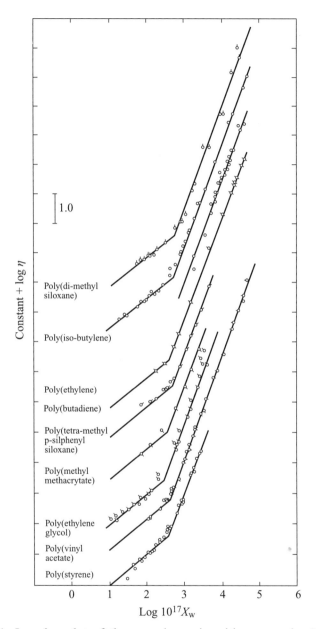

Figure 7.1. Log–log plot of the zero-shear viscosities versus log X_w, where the molecular weight αX_w, for a variety of polymers. The straight lines have slopes of 1.0 or 3.4, and the curves have been arbitrarily shifted along the ordinate (Graessley, 1974).

cosity of liquid polymers upon their molecular weights as evidenced in Figure 7.1. Because $IQ = V_i/V_0$ and V_i increases with molecular weight at a faster rate than does V_0, because V_i depends on the volume pervaded by the polymer coil or $(\langle r^2 \rangle)^{3/2}$ and V_0 depends solely on the polymers molecular weight, it can be shown that IQ depends on the molecular weight to a power higher than $\frac{1}{2}$. As the molecular weight of a polymer increases we can expect the overlap of their random coils or entanglements to increase. M_c may be the minimum molecular weight required to achieve an IQ sufficiently large so that the slow flow of the liquid polymers begins to be dominated by the unraveling of the entanglements between polymer coils.

The value of M_c for a given polymer liquid has important practical consequences for its processing, because not only the viscosity, but also its dependence on molecular weight, is very sensitive to its molecular weight, i.e., whether or not it is $<M_c$ or $>M_c$. Both the energy and time requirements for processing a polymer liquid are affected by whether or not its molecular weight is above or below M_c.

In addition to the molecular weight or chain length, the viscosity of a polymer liquid is also sensitive to the rate of flow or shear rate as evidenced in Figure 7.2. As the shear rate is increased we notice a

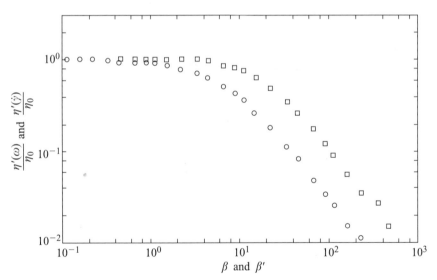

Figure 7.2. Dynamic viscosities versus shear rate observed for two samples of polystyrene: $\overline{M}_w = 215,000$ (○) and $\overline{M}_w = 581,000$ (□).

significant drop in the viscosities of both polymers, indicating that distortion of the polymer coils by the shearing forces reduces the ability of overlapped/entangled polymer coils to retard their flow (both polystyrene samples have molecular weights well above M_c). Also note that a higher shear rate must be applied to the polystyrene sample with the higher molecular weight before its viscosity begins to decrease. This may indicate that the more densely entangled polymer (higher IQ) is more difficult to unentangle and thus requires a higher rate of shear to disrupt the extensive network of overlapped polymer coils, which leads to an increased/decreased flow rate/viscosity.

POLYMER GLASSES

Bulk, amorphous polymer samples must possess sufficient thermal energy or be above a certain temperature before they can flow irreversibly under the application of a steady force such as gravity. Below this temperature, T_g, the sample is rigid and glassy, while above T_g the polymer sample is tough and leathery and, if given sufficient time, may eventually flow. T_g is called the glass-transition temperature and is a fundamental parameter for both the processing and use of polymer samples. Above T_g the polymer coils in the bulk, amorphous sample have the ability to change their conformations and, if provided with sufficient time, may unentangle from the network of overlapped polymer coils and evidence bulk flow. The glass-transition temperature T_g is a manifestation of the ability of an amorphous, bulk polymer sample to flow, but is not characteristic of a change in phase, such as the melting temperature, between a solid and a liquid, because the polymer chains are disordered both above and below T_g. For example, in Figure 7.3 the glass-transition temperature T_g of an amorpous polymer and the melting temperature T_m of a crystalline polymer as monitored by the temperature dependence of the specific volumes of both samples are shown. Note in (a) that the plot of $\overline{V}$ versus T shows a change in slope at $T = T_g$ for the amorphous polymer, while in (b) an abrupt step or increase is seen in $\overline{V}$ at $T = T_m$ for the crystalline polymer.

This qualitatively different behavior is a reflection of the absence/presence of a structural change at T_g/T_m as a consequence of the absence/presence of two distinct phases in equilibrium. At T_g the

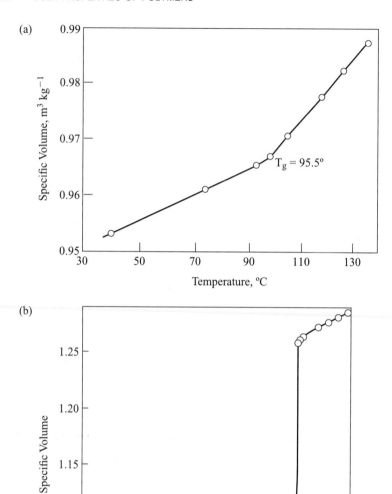

Figure 7.3. (a) The change in the specific volume of atactic polystyrene with temperature (Gordon and Macnab, 1953). (b) The change in the specific volume of polyethylene with temperature (Chiang and Flory, 1961).

amorphous polymer sample remains disordered, but its polymer coils are now able to undergo conformational interconversions and flow, while at T_m the conformationally rigid polymer chains in crystalline sample portions are in equilibrium with conformationally mobile chains in the molten, liquid portions of the sample. Thus T_g reflects a transition in polymer dynamics or motion and not in structure. The glass transition temperatures observed for several different polymers are presented in Table 7.1 and are seen to be strikingly dependent on their chemical microstructures. The glass-transition temperatures of polymers are generally discussed in terms of three factors: (1) the inherent flexibilities/backbone bond rotation barriers of their individual chains, (2) the size or steric bulk of their side chains, and (3) the interactions (steric, dipolar, hydrogen bonding, etc.) between chains. To date, it has not been possible to successfully connect the T_g of polymers with their microstructures even when consideration is given to the above factors. It does seem, however, that among these factors the interactions between polymer chains are most important in establishing the magnitude of their T_g's.

The glass-transition of a polymer may be simply visualized with the the aid of a styrofoam cup made from foamed polystryrene and a hair dryer. Hold the cup with tongs and apply hot air from the hair dryer. The cup will soon begin to distort and crumple as its temperature exceeds the T_g of polystyrene ($\sim 100°C$). Upon removal of the hot air and after cooling, the deformed styrofoam cup will once again become rigid, but now in the deformed shape achieved by exceeding the T_g of polystyrene.

Now let us examine the time-dependent behavior of an amorpous polymer sample above its glass-transition temperature. Figure 7.4(a) presents the modulus (ratio of the stress required to achieve a given strain) of an amorphous polymer sample measured at $T > T_g$, as determined in Figure 7.3, for example, for various times of duration of the applied stress. In general, materials with high modulus are very strong in the sense that large forces or stresses (force/area) are necessary to produce very modest strains or extensions. Notice that for very brief applications of the stress, the modulus of an amorphous polymer sample above its T_g is high and its response is that of a glassy material. As the duration of stress (t) is increased, the modulus decreases, eventually reaching a plateau region, where the modulus remains constant over approximately four orders of magnitude in the

TABLE 7.1. Glass-Transition Temperature for Selected Polymers

Polymer	Repeating Unit	Glass-Transition Temperature (°C)
Silicone rubber	$\begin{array}{c} CH_3 \\ | \\ -Si-O- \\ | \\ CH_3 \end{array}$	-125
Polybutadiene	$-CH_2-CH{=}CH-CH_2-$	-85
Polyisobutylene (butyl rubber)	$\begin{array}{c} CH_3 \\ | \\ -C-CH_2 \\ | \\ CH_3 \end{array}$	-70
Natural rubber	$\begin{array}{c} CH_3 \\ | \\ -CH_2-C{=}CH-CH_2- \end{array}$	-70
Polychloroprene (neoprene rubber)	$\begin{array}{c} Cl \\ | \\ -CH_2-C{=}CH-CH_2- \end{array}$	-50
Poly(vinyl chloride) (P.V.C.)	$\begin{array}{c} Cl \\ | \\ -CH_2-CH- \end{array}$	$+80$
Poly(methyl methacrylate) rubber)	$\begin{array}{c} CH_3 \\ | \\ -CH_2-C- \\ | \\ C{=}O \\ | \\ O \\ | \\ CH_3 \end{array}$	$+100$
Polystyrene	$\begin{array}{c} -CH_2-CH- \\ | \\ C \\ \diagup\ \diagdown \\ HC\ \ CH \\ || \quad | \\ HC\ \ CH \\ \diagdown\ \diagup \\ C \\ | \\ H \end{array}$	$+100$

duration of the applied stress, and the response is that of a rubbery material. Finally, beyond a certain duration of the stress, the modulus drops precipitously, the polymer sample no longer is able to support the stress, and macroscopic flow of the sample begins.

(a)

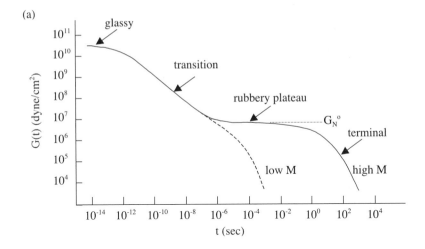

(b)

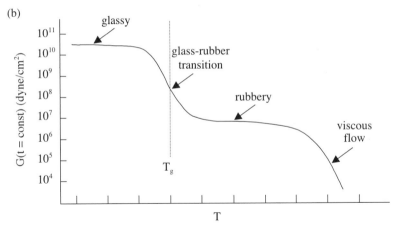

Figure 7.4. Modulus $G(t)$–time plot at $T > T_g$ (a) and modulus $G(t = \text{const})$–temperature plot (b) for a typical amorphous polymer.

Let us discuss the underlying time- or frequency-dependent response of an amorphous, bulk polymer sample observed above its glass-transition temperature. Clearly, if given sufficient time to respond to a force or stress—that is, if the force or stress is applied for a sufficiently long period—the sample will evidence overall, macroscopic flow. We can see from Figure 7.4(a) that the time required to produce flow might be as long as seconds or minutes. Apparently, the time required to unentangle the overlapped polymer coils in an amorphous, bulk sample, to permit displacements of their center of masses, and result in macroscopic flow is substantial. As a consequence, if stresses are applied for shorter times, the sample will be unable to respond by flowing, but instead will behave as an elastic, rubbery material for intermediate times of stress application or will eventually behave as a rigid glassy material for stresses applied for very short times. In the rubbery plateau region the polymer chains have sufficient time to alter their conformations, but not enough time to unentangle, so the entanglements serve as cross-links forming a network that holds the sample together, allowing reversible deformations but not irreversible flow. Note the dotted line in Figure 7.4(a) corresponding to low-molecular-weight, amorphous, bulk samples which show no rubbery plateau response. This behavior is seen when the molecular weight is below the level (M_e) needed to produce a well-developed entanglement network of overlapped polymer coils and is analogous to the molecular weight M_c that characterizes the change in the molecular weight dependence of melt viscosity from the first to the 3.4 power, though M_c and M_e are not necessarily identical. When stressed for very short times, the response becomes glassy and brittle, because time is now insufficient to even permit alterations in the conformations of individual polymer coils.

If we apply a constant sustained stress on an amorphous, bulk polymer sample and observe its deformation or strain as a function of temperature, then the resulting modulus (stress/strain) shows the behavior depicted in Figure 7.4(b). Under a constant load our amorphous polymer sample exhibits first rigid, glassy behavior at low temperatures, then at a temperature that can be identified with T_g a transition from glassy to rubbery behavior, and at $T \gg T_g$ our sample begins to flow. Notice the remarkable similarity between the curves in parts (a) and (b) of Figure 7.4. This equivalence between the time- and temperature-dependent mechanical responses of amorphous, bulk polymers is often called the time–temperature superposition principle. In other words, amorphous bulk polymers respond equivalently to the time, or inversely to the frequency, dependence of stress application and to the temperature when under constant stress (Ferry, 1980). This behavior is characteristic of polymers and is the basis for many of their unique applications.

Clearly it is not sufficient to describe the T_g of an amorphous polymer in order to predict its response to a stress. In addition, we must know the frequency of the applied stress or the duration of its application. This fundamental feature of the behavior of materials made from long-chain polymers or macromolecules can be readily visualized with a simple demonstration called the "Slime Experiment" (Casassa et al., 1986). Mixing a 5 wt% aqueous solution of poly(vinyl alcohol) (PVOH) with a small amount of a 5 wt% aqueous solution of sodium borate (borax) leads to the production of a very viscous mixture. The borate ions interact with the hydroxyl groups on the PVOH chains, creating a highly swollen, temporarily cross-linked network often called "slime." The slime will flow if given sufficient time, because the –OH‑‑‑borate‑‑‑HO– cross-links are only physical in nature and will break and reform, producing a highly viscous material. In fact, if a portion of the slime is placed on the edge of a desk and pulled slightly over the edge, eventually the slime will flow to the floor without breaking; that is, a continuous strand of slime will result. This behavior is characteristic of a viscoelastic liquid, because it exhibits irreversible macroscopic flow under the influence of gravity. On the other hand, if the slime is rolled up into a ball and dropped from a height of 2–3 feet onto a desk top, the slime ball will bounce elastically several times, much like a typical rubber ball, but will soon stick to the desk surface and slowly flow and spread. In this instance the slime behavior is at first characteristic of an elastic solid as it bounces, but later after sticking to the desk top it appears to be a viscous liquid and once again flows irreversibly. Finally if we roll out the slime sample into an elongated cylinder and pull sharply on both ends, the cylinder of slime will fracture. Examination of the fractured sample reveals clean, sharp, well-defined fracture surfaces reminicent of those observed, for example, when an icicle is snapped in two. Slime's response to a briefly applied, large force is to fail as a brittle, glassy solid. Remember that our slime essentially consists of 95% water and 5% PVOH, with a much smaller amount of $Na_2B_2O_3$.

These slime demonstrations evidence behavior closely parallel to that presented in Figure 7.4(a) for a bulk, amorphous polymer sample above its glass-transition temperature. Given sufficient time to respond, both the slime and the bulk, amorphous polymer samples behave as viscous liquids and flow irreversibly. At shorter times both samples respond elastically, and finally for very short response times each behaves as a rigid, glassy material, exhibiting very little deformation in response to a large force. In summary, both the highly swollen slime sample and the undiluted, amorphous polymer sample exhibit mechanical responses characteristic of a low modulus liquid, a moderate modulus elastic solid, and a high modulus glassy solid, depending on the amount of response time given to each sample. How can we understand the qualitatively comparable mechanical behavior exhibited by a bulk, amorphous, highly entangled polymer sample and the highly swollen slime? Both samples contain mobile, disordered, randomly coiling polymer chains, albeit at drastically different concentrations. Nevertheless, even at 5% concentration in slime the PVOH chain coils are extensively overlapped, but at a level much reduced from that in the bulk, amorphous polymer sample. Disregarding the presence and role-played by the $Na_2B_2O_3$ in slime for a moment, we would expect a much-reduced degree of entanglement in this highly swollen sample, because of the much greater distance between the centers of mass of the PVOH coils.

The key to this puzzle lies with the borax ($Na_2B_2O_3$). Though not as extensively overlapped as the polymer coils in the amorphous, bulk sample, where the PVOH coils do overlap they interact much more strongly via the physical attraction between pendant $-OH$ groups that is mediated and strengthened by the bridging borate groups [i.e., $\sim OH\text{---}(B_2O_4)^{-2}\text{---}HO\sim$] which behave as time-dependent, weak cross-links between the PVOH coils. Thus, though smaller in number, the more dilute PVOH cross-links in slime are more effective or stronger than the more numerous entanglements in the bulk polymer sample, leading to qualitatively similar time-dependent mechanical responses.

POLYMER ELASTOMERS

Let us now permanently tie together the polymer coils in an amorphous, bulk polymer sample by introducing chemical cross-links between a few of the repeat units belonging to different chains. This procedure yields a cross-linked polymer network, and the first such example was discovered by Charles Goodyear well over 150 years ago. After much trial-and-error experimentation, Goodyear found that heating natural rubber, poly-*cis*-1,4-isoprene, with sulfur produced a completely elastic material that, unlike raw natural rubber,

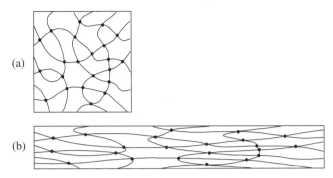

Figure 7.5. Cross-linked network in (a) an unstrained state and (b) a strained state.

evidenced no permanent plastic deformation. Later it was shown (Farmer and Shipley, 1946; Naylor, 1946; Bloomfield, 1946) that the sulfur produced chemical cross-links between the rubber polymer chains, as indicated below, leading to a cross-linked polymer network as illustrated in Figure 7.5. At sufficiently high degrees of cross-linking, each rubber polymer chain is chemically attached to the network, creating in essence a single macroscopic molecular network.

$$
\text{Poly-}cis\text{-1,4-isoprene} \quad -(CH=\overset{\overset{\displaystyle CH_3}{|}}{C}-CH_2-CH_2-)_n-
$$

$$
\cdots\cdots\cdots\cdots-CH_2-\overset{\overset{\displaystyle CH_3}{|}}{\underset{\underset{\displaystyle S_x\ (x<6)}{|}}{C}}-CH=CH-CH_2-\cdots\cdots\cdots\cdots\cdots
$$

$$
\cdots\cdots\cdots\cdots-CH_2-\underset{\underset{\displaystyle CH_3}{|}}{C}-CH=CH-CH_2\cdots\cdots\cdots\cdots\cdots
$$

As indicated in Figure 7.5(b), the cross-linked sample (above T_g) may be extended by application of a force, but instead of retaining its deformed shape after removal of the force, it is observed to return to its original underformed shape as in Figure 7.5(a). Introduction of the cross-links to create a polymer network has produced an amorphous polymer sample with a memory of its shape during the cross-linking process. We observe this phenomenon daily when we take a rubber band and stretch it around an object like a newspaper, to hold its pages together, and then remove the rubber band to read the newspaper. After removal, the used rubber band appears indistinguishible in form/shape/size from when it was first removed from a box of rubber bands and before it was used to secure the newspaper. Depending on the number of pages contained in the newspaper, the rubber band may be required to stretch several times its undeformed length, as indicated in Figure 7.5, in order to encompass and hold together the newspaper. The force required to achieve this large extension of a rubber band is quite minimal, when compared to the forces required to deform other solid materials. The modulus of the rubber band—that is, the stress required to produce a given strain—is small when compared to typical solids composed of small molecules or atoms. For example, a steel wire 1 mm in diameter may be elastically extended by 1% when a force of 1600 N (about twice the weight of a man) is applied. A rubber band of the same diameter maybe stretched 1% by application of a force less than 10^{-2} N. Consequently, the modulus (ratio of stress to strain) of steel is more than 100,000 times that of a rubber band (Treloar, 1975).

The preceding discussion may be summarized by saying that cross-linked polymer samples observed at $T > T_g$ exhibit high extensibility that is reversible and characterized by a very low modulus. What is it about a network of randomly coiled polymer chains that leads to this unique behavior? Once again, it is their long-chain character that permits polymers to assume a vast array of sizes and shapes by simply rotating about their backbone bonds. When a cross-linked polymer is stretched, the coiled chains between network cross-links are also stretched (see Figure 7.5) and assume more extended conformations. The capacity for stretching individual network chains can be demonstrated with the help of Figure 7.6, where a short 8-bond chain and a long 1000-bond chain are presented. Even in the case of the short chain, the distance between its ends can be substantially altered by changing the conformations of its intervening bonds. This is reinforced in Figure 7.6(c), where substantial extension of the 1000-bond chain can be readily envisioned by altering its conformation.

It is clear that individual polymer coils between the cross-link junctions in a polymer network are readily extensible, but why does the network return to its undeformed shape when the extension force is removed? Without assuming a deep knowledge of thermodynamics (Castellan, 1983), it can still be said that any thermodynamic system

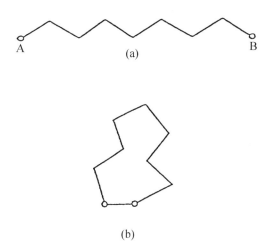

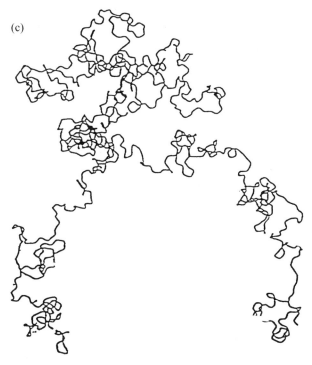

Figure 7.6. (a, b) Extended, compact conformations of an 8-bond chain segment. (c) Polyethylene chain containing 1000 freely rotating bonds.

such as a rubber band always adopts or rearranges itself to achieve a state of low energy and high entropy. Unlike energy, a few words should be offered concerning the physical nature of entropy. Entropy is a measure of the disorder or number of different arrangements of molecules or choices available to a system. Each disordered state, molecular arrangement, or choice i is weighted exponentialy according to the Boltzmann factor of its energy, that is, $\exp[-E_i/RT]$. In other words, a thermodynamic system that has available a large number of disordered states, molecular arrangements, or choices, each associated with a comparable energy, will also possess a large entropy (Hill, 1962).

Now let us consider both the energies and entropies of a cross-linked polymer network at $T > T_g$ (a rubber band, for example) at rest and after stretching. Experiments repeatedly demonstrate that the total volume of a rubber band remains constant during stretching, so the average distance between the polymer coils is unaffected by stretching (Flory, 1953; Treloar, 1975). This suggests that contributions to both the energy and entropy of the rubber band made by interchain or intermolecular interactions are negligibly affected by stretching. In fact, as indicated in Figure 7.5, if in the stretch direction the rubber band is lengthened by a factor of L/L_0, then the other two dimensions contract by a factor of $1/\sqrt{(L/L_0)}$, so the total rubber-band volume remains constant. On the other hand, the intrachain contributions made to the energy and entropy of the rubber band are usually negligible and considerable, respectively. We can see in Figure 7.7 that the distance between the ends of a randomly coiling polymer chain can be greatly altered solely by rotating about a single backbone bond (Abe and Flory, 1970). Because there are generally several hundred backbone bonds in the polymer chain between cross-links, changing the conformation of a single or even a few bonds as the chain is stretched would not be expected to seriously effect the energy of the rubber band.

However, the internal or conformational entropies of the polymer chains between cross-links are seriously impacted by their stretching induced extension. This is made apparent by Figure 7.6, where in parts (a) and (b) different conformations of the same 8-bond chain are presented. Note that in (a) the fully extended length can only be achieved by adoption of the all trans, planar zigzag conformation. In (b), however, the same separation between the chain ends can be obtained by conformations in addition to those indicated there. In

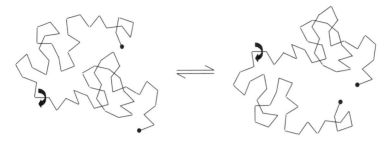

Figure 7.7. Illustration of the reduction in the end-to end separation of a polymer chain end achieved by altering the conformation about a single backbone bond.

general, as a randomly coiling polymer chain is extended the total number of conformations available to it decreases, so its conformational entropy decreases. It is this decrease in the conformational entropies of the network chains, which accompanies the stretching of a rubber network, that both resists stretching and causes the network to snap back to its original dimensions. The retractive elastic force exerted by a cross-linked polymer network is generally almost exclusively entropic in origin (Flory, 1953; Treloar, 1975).

Figure 7.8 presents a schematic repesentation of the responses of

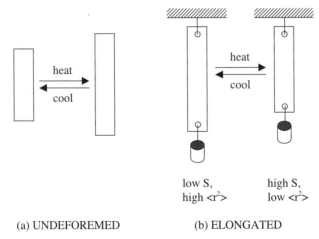

(a) UNDEFOREMED   (b) ELONGATED

Figure 7.8. Length changes in rubber samples upon heating and cooling. In (b) the deforming force is suggested by the weight hanging from the rubber band.

an elastic rubber sample to heating and cooling when the sample is undeformed (a) and under tension and elongated (b). Note like most solids the undeformed rubber sample expands on heating and contracts when cooled. However, when strained the rubber sample contracts when heated and expands when cooled. After stating that the degree of disorder in a thermodynamic system generally increases with an increase in system temperature, I leave it to the reader who now understands the origin of the restoring force in strained rubber networks to explain the behavior depicted in Figure 7.8(b) and to explain the observations made in the rubber-band demonstration presented in Chapter 1.

The retraction of a rubber elastic polymer network under tension produced by warming the network may be easily demonstrated. Cut a rubber band and tie one end to a paper clip and the other end to a long string. Slip the string and rubber band inside an ~ 1 cm diameter glass tube until the paper clip abuts the top of the tube and the string protrudes from the bottom of the tube. Tie a small weight (25–50 g) to the protruding string, thereby placing the rubber band under tension. With a hair dryer carefully heat the glass tube, and consequently the rubber band. Using a yard stick as a measure, you will note that upon heating the distance between the bottom of the glass tube and the suspended weight will decrease, because the ruber band has contracted upon being heated.

Let us now contrast this behavior with that generally exhibited by molecular and atomic solids when they are deformed. Because solids composed of atoms and/or small molecules do not have the ability to change there sizes and shapes, unlike polymeric materials they do not have an internal means for responding to an external stress. Instead, they can only respond interatomically or intermolecularily by rearranging their spatial distributions. For elastic deformations this means that a change in sample volume is necessary, and so the resistance to deformation becomes primarily energetic in origin as the atoms or molecules are moved away from their equilibrium distances of separation and forced into a new spatial arrangement by the stress. A similar situation occurs even for a cross-linked polymer network when it is stressed at $T < T_g$, because now the polymer chains between cross-link junctions are unable to respond internally by changing their conformations. In this instance the glassy polymer network must also respond by adjusting the intermolecular spatial distribution of polymer segments, without benefit of intramolecular conformational adjustments. Here too the deformation is accompanied by a volume cha· ~~re~~ and is principally resisted by an increase in the energy of the strained system.

To demonstrate that conformational mobility of the constituent polymer network chains is necessary for reversible elastic behavior to be exhibited by a cross-linked polymer network, we may utilize a rubber band, a pair of large tongs, and a beaker of liquid nitrogen. Place the rubber band around the tongs and stretch the rubber band by opening the tongs. Place the tongs and stretched rubber band in the liquid nitrogen for a few seconds, and then remove and close them to allow the cold, stretched rubber band to fall off and rest on a paper towel. Note that after the rubber band warms it slowly contracts to its original size, because its T_g has been exceeded allowing the network chains the conformational mobilty necessary to contract to their original, more compact, higher entropy conformations. Natural rubber, cis-1,4-polyisoprene, has a T_g of $\sim -70°C$ (See Table 7.1), so the liquid nitrogen was sufficiently cold to depress the temperature of the rubber band below its T_g, thereby preventing its retraction via the interconversion of the network chains from their extended to their undeformed compact conformations.

In addition, if the rubber band is struck with a hammer immediately upon its removal from the liquid nitrogen, it will shatter like a glassy, brittle solid, because it is well below its T_g and its constituent network chains cannot respond internally by altering their conformations.

We can see that amorphous polymer solids, whether cross-linked or not, are versatile materials whose mechanical responses depend on and can be tuned to their use temperature. Whereas a cross-linked polystyrene $(T_g = 100°C)$ may be hard and brittle at room temperature, it may function very well as an elastic solid at temperatures above $100°C$ and be used, for example, as a material for high-temperature flexible hoses and gaskets.

Once an amorphous bulk polymer sample is cross-linked, its macroscopic dimensions cannot be irreversibly altered without destroying some of the bonds in the polymer backbones and or in the cross-links. In this sense it resembles a thermoset polymer like the phenol-formaldehyde and urea-formaldehyde, step-growth polymers mentioned in Chapter 2. Once cross-linked, the network polymer sample cannot be reformed. However, unlike the highly cross-linked thermosets, elastic polymer networks may be swollen upon exposure to liquids that are good solvents for the un-cross-linked polymer. The network polymer can absorb large quantities of solvents and exhibit swollen volumes many times their dry volume. (This can be readily demonstrated by placing a rubber band in a beaker of toluene.) Of course in a swollen polymer network the constituent chains are stretched and under tension. When this elastic restoring force is balanced by the swelling force due to the enthalpy and entropy of mixing the solvent molecules and the polymer segments in the swollen network, the swollen network becomes saturated with the swelling liquid and is unable to undergo further swelling. The degree of swelling experienced by an elastic polymer network depends on the quality of the solvent for the polymer and on the degree of cross-linking or the average number of backbone bonds between cross-links in the network (Flory, 1953). The higher the density of cross-links, the lower the equillibrium degree of swelling, and this behavior can be used to estimate the degree of network cross-linking.

The ability of polymer networks to swell when exposed to solvents is taken advantage of in several applications. Soft contact lenses and highly absorbant layers in disposable diapers are two among a host of applications utilizing the ability of cross-linked polymers to be significantly swollen with a variety of liquids.

CRYSTALLINE POLYMERS

Having discussed the behaviors of amorphous, bulk polymers above and below their glass-transition temperatures, for both uncross-linked and network samples, we turn finally to bulk polymer samples possessing crystallinity. In Figure 7.9 we see the electron microscopic images of a polyethylene sample crystallized from solution and the melt. Note the flat lozenge shape of these lamellar crystals. Electron diffraction studies of these and many other polymer crystals demonstrated (Storks, 1938; Keller, 1957) that the extended crystallized chains are arranged in the direction perpendicular to the large faces of these lozenge-shaped crystals. Because the highly extended crystalline chains are much longer than the thickness of these lamellar crystals (typically 100–200 Å)—for example, a 10,000-bond polyethylene chain has a length of more than 12,000 Å in its fully extended, all-trans crystalline conformation (see Chapter 5)—the polymer chains must

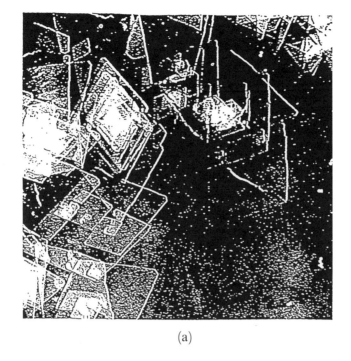

(a)

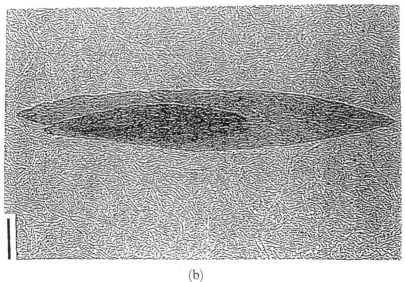

(b)

Figure 7.9. Single crystals of polyethylene obtained from solution (a) and melt (b) crystallization.

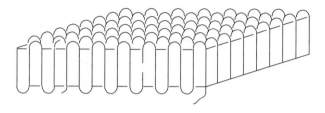

Regular, adjacent re-entry folds similar to those postulated
as present in pyramidal crystals that have been grown from
solution.

(a)

Switchboard, or nonadjacent re-entry model in which a
nonordered amorphous layer is present on both sides of the
lamellae.

(b)

Figure 7.10. A schematic representation of a lamellar polymer crystal drawn
assuming regular chain folding (a) and irregular chain folding (b) (Flory,
1962).

fold, as indicated in Figure 7.10, in order to be accommodated. The
folds in Figure 7.10(a) have been depicted as being uniformly tight
with each reentering the crystalline surface in the position adjacent to
where it exited. This crystalline morphology is usually termed regular,
adjacent reentry chain-folding and is contrasted to the situation in (b),
called the switch-board model, where the folds are irregular in length
and reenter the crystal surface at random positions. In general, depending on conditions of crystallization, such as temperature and rate
of crystallization and concentration of polymer in the crystallizing
medium, the surface of lamellar polymer crystals are expected to be
organized closer to or farther from either of the two extreme morphologies depicted in Figure 7.10.

As an example, when crystallized from the melt, a spherulitic morphology of the type seen in Figure 7.11 is generally observed (a), where the underlying structure of the spherulites is drawn in (b). Note that even in rapid crystallization from the melt the fundamental structural feature observed is the chain-folded, crystalline lamella. Below the lamellar surface, the detailed crystal structure of polyethylene is presented in Figure 7.12, where each chain is fully extended in the all-*trans* conformation. As mentioned previously, the IQ $= V_i/V_0$ of a polymer chain in its crystal approaches 1; that is, the volumes physically occupied and influenced are nearly coincident. This means that each chain is interacting with only a small number of nearest-neighboring chains; however, this interaction extends over a large distance on the order of the lamellar thickness of 100–200 Å. At the same time, those portions of the polymer chains on and between the fold surfaces in the interlamellar regions are disordered and would be expected to have IQs and behaviors similar to amorphous, bulk polymers. To a first approximation then, we expect polymers that crystallize to be composed of two phases: intralamellar crystalline and interlamellar amorphous phases.

The crystalline regions in bulk polymer samples serve to anchor the interconnecting amorphous regions, thereby serving as ordered reinforcing elements embedded in an amorphous matrix and leading to the notion that crystalline polymers, which are really semicrystalline, can be considered single-component, composite materials. Crystalline polymers generally exhibit strengths that are much improved over bulk, amorphous samples. If the crystalline domains can be aligned and the chains in the amorphous regions extended both along a single direction, say the long axis of a fiber, then a material extremely strong in the fiber direction results. Commercial examples can be found in Kevlar and Spectra fibers, which are highly aligned and drawn poly(paraphenylene terephthalmide) and polyethylene fibers, respectively, and are much stronger than steel on a per-unit weight basis. Consequently, these two polymer materials are replacing metals in applications that require both high strength and low weight, such as high-performance tire cord and cabeling and armor, both personal and as cladding on aircraft and tanks.

Of course the overall mechanical behavior of a semicrystalline polymer sample depends on the mobility of those portions of the chains in the amorphous regions of the sample. At $T > T_g$ the amor-

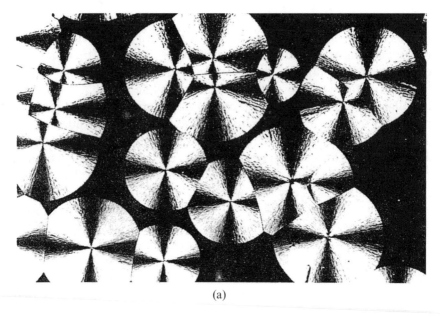

(a)

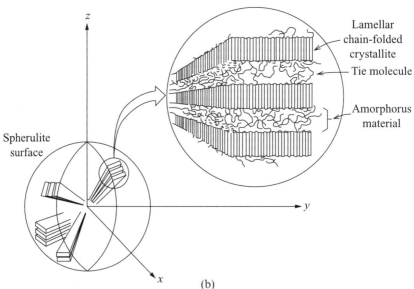

(b)

Figure 7.11. (a) Spherulitic growth in a melt of isotactic polystyrene (115×) (Bovey, 1979). (b) Schematic representation of the detailed structure of a spherulite as determined by small-angle X-ray scattering (Boyd and Coburn, as cited in Mandelkern, 1972). Crystallization generally continues until spherulites have grown and abut their neighbors.

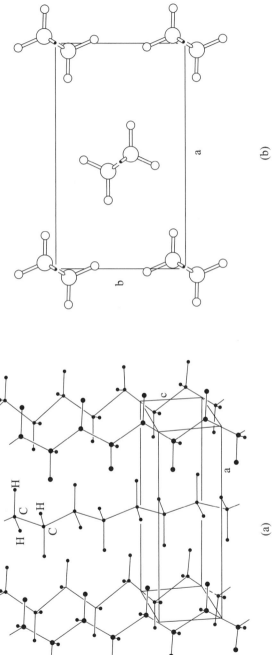

Figure 7.12. Crystal structure of polyethylene: (a) General view and (b) projection along the *c*-axis (chain direction) as viewed looking down on the lamella below the fold surface.

phous chains are able to change their conformations in response to a stress, but macroscopic flow is prevented by the reinforcing crystalline domains. This leads to a material that is strong and tough, but not brittle, because the crystalline domains act as macroscopic cross-links keeping the flexible, amorphous chains from plastically deforming. The response of a semicrystalline polymer sample is similar to that of an amorphous sample below T_g, except that the presence of the crystalline domains leads to improved strength. The response both above and below T_g may be substantially modified by altering the sample morphology, including the degree of crystallinity, crystallite size, the alignment of crystallites, and the extension of chains in the amorphous domains. Consequently, the final use properties of semicrystalline polymer samples are highly dependent upon the means used to process them. This provides the materials engineer with the ability to tailor the properties of a semicrystalline polymer material by processing alone, without recourse to employing a chemically distinct polymer to satisfy the requirements of a particular application. For example, polyethylene is familiarly used to make lightweight and flexible, yet only moderately sturdy, trash bags. However, polyethylene can also be processed into the strongest man-made fibers known, which are used in the lightweight armor for the most advanced stealth aircraft.

MELTING OF POLYMER CRYSTALS

More important to the physical properties of semicrystalline polymers than the T_g of the amorphous regions is the melting temperature T_m of the crystalline lamellae. Unlike T_g, which signals a change in the mobility, but not the order, in an amorphous, bulk polymer, at T_m the rigidly ordered polymer chains in the crystal (see Figure 7.12) melt into a disordered liquid of conformationally mobile polymer chains able to flow macroscopically. Though the physical response of a semicrystalline polymer depends on whether or not T is greater than T_g, its macroscopic dimensions do not. However, above T_m the sample is no longer able to support long-term stresses, and so it deforms plastically and flows. Table 7.2 presents melting temperatures of several polymers, where the tremendous range in T_m's is plainly evident. From silicone rubber [poly(dimethyl siloxane)] with a $T_m = -40°C$ to the slippery Teflon [poly(tetrafluoroethylene)] with a $T_m = 330°C$, the melting

TABLE 7.2. Melting Temperature of Selected Crystalline Polymers

Polymer	T_m, °C
Poly(dimethyl siloxane)	−40
cis-1,4-Polybutadiene	6
cis-1,4-Polyisoprene	36
Poly(ethylene oxide)	66
trans-1,4-Polyisoprene	74
Polyethylene	140
trans-1,4-Polybutadiene	148
Polystyrene (isotactic)	240
Nylon-6,6	267
Poly(ethylene terephthalate)	270
Poly(acrylonitrile)	300
Poly(tetrafluoroethylene)	330

temperatures of polymers define the upper temperature limits for their use in applications requiring dimensional stability. Though we discuss this topic further later in the chapter, for now simple mention is made of the two factors that dominate the melting of polymer crystals.

The first is the differences in the interactions between polymer chains in the crystal and in the expanded, lower density melt. Not only are the chains farther apart in the melt, but their IQ is many times larger than for the crystalline chains (IQ ∼ 1), which, however, interact over much longer portions of their contours (the lamellar thickness ∼100–200 Å) and more closely with the small number of immediately neighboring chains in their crystalline unit cell (see Figure 7.12). Thus, energy must be expended to melt the crystals, with the major requirement originating from the changes in interchain interactions and density, which accompany the transition. In the melt, polymer chains are disordered and are free to adopt their full complement of conformations, while in the crystal the chains are generally restricted to a single, highly extended conformation. Because polymers almost invariably adopt the conformation of lowest intramolecular energy when crystallized, very little of the energy required to melt a crystalline polymer arises from the intrachain process of conformational disordering, but instead is necessary to overcome the interchain interactions which are stronger in the closely packed crystals.

At the same time that energetically favorable interactions between polymer chains are reduced in the melt, the disorder due to volume expansion (decrease in density) and the ability of the molten polymer chains to adopt a plethora of conformations subtantially increases their entropy. The intrachain conformational freedom gained by melting crystalline polymers dominates the entropy increase generated by the increased volume available to the melt. These considerations lead therefore to the following qualitative principles governing the melting temperatures of crystalline polymers: (1) Polymers with strong interchain interactions, such as hydrogen-bonding and dipolar interactions, will generally have higher T_m's compared with nonpolar polymers limited to van der Waals interactions, and (2) flexible polymers able to adopt more randomly coiling conformations in the melt will generally melt at a lower temperature than stiff, conformationally less diverse polymers. Thus polar, hydrogen-bonding polymers that are conformationally limited or stiff when molten will have high T_m's, while nonpolar, conformationally flexible polymers will have low T_m's.

POLYMER FIBERS

Conversion of semicrystalline polymers into fibers, in addition to its commercial significance, serves as an important illustration of how the physical properties of polymeric materials can be significantly influenced by the means used to process them. Schematic diagrams of fiber spinning are presented in Figure 7.13. In melt spinning the molten fibers emerging from the spinnerette are cooled and solidified and then drawn. When spinning from solution the fibers are solidified either by flashing off the solvent (dry spinning) or by passage through a coagulation bath (wet spinning) containing a nonsolvent for the polymer. After extrusion of the polymer melt or solution through the fine diameter holes of the spinnerette and consolidation of the fiber, most fibers are subjected to a drawing process, whereby they are wound up at a speed greater than they emerge from the spinnerette. This not only reduces the fiber diameter, but is believed to also align the crystalline regions along the fiber axis, extend the amorphous portions of polymer chains and align them along the fiber axis, and increase the crystallinity of the fiber.

Figure 7.14 illustrates the morphological changes produced by fiber

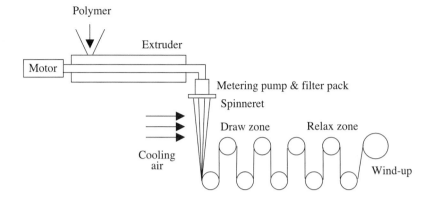

(a) Melt spinning

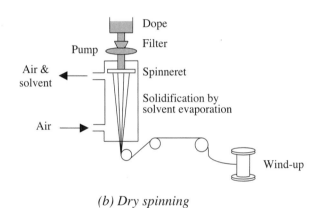

(b) Dry spinning

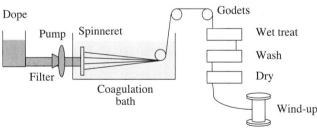

(c) Wet spinning

Figure 7.13. Schematic of (a) melt, (b) dry, and (c) wet spinning of fibers.

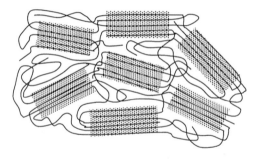

Fringed micelle model (Crystalline regions are shaded.).

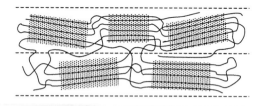

Fringed fibril model of drawn fiber structure.

Figure 7.14. Morphological effect of fiber drawing.

drawing, which tends to align both crystalline and amorphous domains and extend the chains in the latter regions. After drawing, a fiber is not only macroscopically anisotropic, with a huge L/d ratio, but its morphology is also highly anisotropic. Through spinning and drawing we seek to align and extend all polymer chains along the fiber axis, so extension of the fiber requires breaking the stretched amorphous chains and/or melting of the crystals. As a result, when done properly, fibers exhibit high moduli in the fiber direction. Their moduli can be increased by orders of magnitude compared with the moduli of isotropic polymer samples. For instance, Kevlar [poly(paraphenylene terephthalmide)] and Spectra (polyethylene) fibers show remarkable extensional strengths, because of the nearly perfect alignment of their crystals and the high extension of their amorphous chains achieved in their processing. They are used as lightweight replacements in applications requiring great strength, such as armor, tire cord, and fiber reinforcements in composite materials.

Because of the microscopically anisotropic arrangement of crystalline and amorphous regions achieved in the processing of high-strength fibers, such as Spectra and Kevlar, their physical responses are also anisotropic. In the fiber direction they have very high moduli as a consequence of the alignment of crystallites and the extension of intercrystalline chains. However, in the radial direction these fibers are relatively weak as evidenced by their low compressive strengths. Figure 7.14 makes evident that, unlike axial deformations, rupture of aligned intercrystalline chains and/or melting of crystalline domains do not necessarily accompany radial deformations of a polymer fiber, thereby producing their characteristically anisotropic mechanical responses. With the exception of liquid crystals, small-molecule and atomic solids, whose constituents do not possess the inherent, individual anisotropy of long-chain polymers, cannot be macroscopically processed into materials with highly anisotropic physical properties. It is the *inside* characteristic of long polymer chains which enables them to be processed into bulk samples whose *outside* responses may be highly anisotropic.

In the remainder of this chapter we briefly revisit some of the behaviors observed for bulk polymers in the hope of establishing closer connections with the microstructures and conformational properties of their individual, constituent chains.

MELT VISCOSITY

Let us begin with the interesting melt viscosity behavior seen in Figure 7.1, where the molecular-weight dependence of the zero shear viscosity is seen to change rather abruptly at M_c from a linear dependence to one proportional to the 3.4 power of the molecular weight. M_c and, more importantly, the number of backbone bonds corresponding to M_c, $N_c = M_c/M_b$, where M_b is the molecular weight per backbone bond, depend on polymer microstructure.

Hence, there is a molecular weight, M_c, for each polymer above which the zero-shear viscosity rises more rapidly with molecular weight than it does at molecular weights below M_c. This rapid rise in the viscosity with molecular weight is caused by entanglements, which are topological constraints on the molecular motion due to the obvious fact that chains cannot pass through each other. Because of the constraints due to entanglements, a long polymer molecule surrounded by other like polymer chains cannot move far in directions perpen-

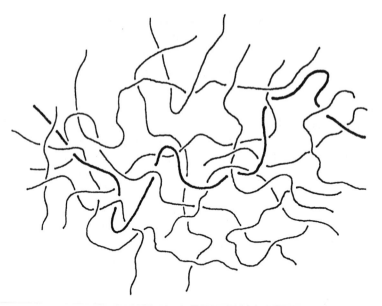

Figure 7.15. A polymer chain and its surroundings in an entangled polymer liquid.

dicular to its own chain contour (Edwards, 1967). Therefore, molecular motion is limited to motions along the contour of the polymer chain. Such a relatively slow, snake-like motion is called reptation (deGennes, 1971, 1979, 1983). Observations of the molecular motions of long DNA molecules entangled with other DNA chains in solution provide a dramatic illustration of this general kind of motion that is unique to polymers and is known as reptation (Perkins et al., 1994).

de Gennes (1971) considered the motion of an individual polymer chain as it moves through a sea of other polymer chains. This situation is analogous to the motion of a snake through a set of obstacles. A truly physical example of a reptating chain can be found in the study of Garter snakes moving through a maze of fixed obstacles (Gans, 1970). It was found that as the density of fixed obstacles increased the force exerted by the snake to move forward, along the length of its body contour, increased rather dramatically.

Based on the reptation model for the motion of a polymer chain in its melt, Doi and Edwards (1978) developed a theory for the dynamics of entangled polymer chains and showed that the zero shear viscosity

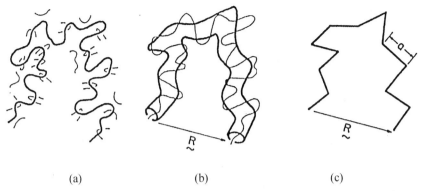

     (a)   (b)   (c)

Figure 7.16. Representations of a polymer chain and its surroundings in an amorphous bulk sample. (a) The chain and segments of neighboring chains, (b) the chain in a tube of uncrossable constraints provided by its neighbors, and (c) the primitive path of a chain among the surrounding constraints provided by neighbors (a = step length of the primitive path; $\mathbf{R}$ = end-to-end vector of the chain) (Graessley, 1982).

scales as the cube of M_w. Subsequent modifications to this theory (Marruci and deCindo, 1980; Marruci and Hermans, 1981) lead to the predictuion that η_0 scales as $M^{3.5}$, in close agreement with experiments (Graessley, 1982), for $M > M_c$. Below M_c we expect $\eta \alpha M$, because without entanglements the friction factor for each polymer should be the simple sum of friction factors for each of its segments, which is proportional to the number of its segments or the molecular weight M (Bueche, 1962).

We may uncover the microstructural dependence of M_c or N_c by considering at what chain length or molecular weight the bulk polymer becomes entangled (Lin and Juang, 1999). Consider a molten sample containing polymer chains with n backbone bonds of length l, whose molecular weight M equals nM_b, where M_b is the average molecular weight per backbone bond. V_0, the volume physically occupied by each polymer, is $M/\rho N_A$, where ρ is the polymer melt density and N_A is Avogadro's number. The volume spanned or pervaded or influenced by each randomly coiling polymer chain can be approximated as $V_i = (4\pi/3)(\langle \mathbf{r}^2 \rangle_0/2)^{3/2}$. Because $C_n = \langle \mathbf{r}^2 \rangle_0/nl^2$, we have $\langle \mathbf{r}^2 \rangle_0 = C_n(nl^2) = (C_n M l^2)/M_b$. The number of polymer chains of length n (or M) that would completely fill or occupy the volume spanned or pervaded or influenced by a single polymer chain of the same length is simply given by $IQ = V_i/V_0 = (4\pi/3)[(C_nMl^2)/2M_b]^{3/2}(\rho N_A/M)$. As mentioned before, IQ should account for the number of chains in contact with any other single chain in a polymer melt. IQ depends only on the size ($\langle \mathbf{r}^2 \rangle_0$) of the polymer coil (i.e., the volume spanned by the chain) and its bulk density ρ in the melt (i.e., that portion of the volume spanned or influenced by the polymer coil that is physically precluded to other chains). Thus, $IQ = 1.481\rho N_A(l^3)(M^{1/2})(C_n/M_b)^{3/2}$.

For example, in a polyethylene melt with $M = 140,000$ ($n = 10,000$), we have $M_b = 14$, $l = 1.53$ Å, $\rho = 0.784$ g/cm^3, $C_n = 6.8$, and $IQ = 113$, or each chain is in contact with more than 100 chains. If we knew how many chains each polymer coil must contact to establish an entangled network that restricts its movements to reptations, we could solve the expression for IQ and obtain an estimate for M_c. Experimentally $M_c = 3800$, and this leads to an IQ on the order of 10. When IQ is estimated for other polymers from their measured M_c, $\langle \mathbf{r}^2 \rangle_0$ or C_n, and ρ, a similar value is obtained [see Fetters et al. (1994) for an analogous development connecting the microstructurally sensitive size ($\langle \mathbf{r}^2 \rangle_0$) of a polymer to the dynamic behavior of its bulk liquid]. This approach enables us to explain the abrupt change in the flow behavior of polymer melts that occurs when $M > M_c$ based on the conformational characteristics of single, isolated polymers that are sensitive to their microstructures. In a sense this means that we may distinguish between the bulk flow behaviors of various polymers based on their microstructures, as a herpetologist might predict the behavior of various snakes based on their anatomical characteristics, such as, for example, their length, girth, length-to-girth ratio, and weight.

It is worth briefly mentioning a recent investigation (Rottach et al., 1999) of the diffusion of simple penetrants in polymer melts. The movement or diffusion of small molecules through polymer melts was modeled by a molecular dynamics simulation performed on a collection of freely jointed polymer chains containing a few small molecules. The behavior observed was qualitatively similar to that expected for diffusion through small-molecule liquids. Only when the backbone bond valence angles of the model polymer chains were constrained to adopt a nearly fixed value, thereby converting them from freely jointed to freely rotating chains, did the modeled diffusive behavior exhibit the characteristics observed experimentally in the diffusion of small molecules through real polymer melts. This example nicely illustrates the potential danger in adopting artificial conformational models for polymer chains, which cannot only lead to an inability to distinguish the behavior of different polymers, as characterized by polymer chemistry, but may also result in a loss or inability to understand the physics of polymers, which is needed to comprehend the unique behaviors that separate them from all other materials.

POLYMER GLASS-TRANSITION TEMPERATURES

Though we have already mentioned that attempts to understand the microstructural sensitivity of the T_g of polymers (see Table 7.1) have not generally been successful, the dependence of the T_g's of copolymers on their comonomer sequence distributions has proven amenable to interpretation based on the measure of copolymer chain flexibilty provided by their conformational entropies. For example, in Figure 7.17(a) the glass-transition temperatures measured (Hirooka and Kato, 1974) for a series of vinylidene chloride-methacrylate copolymers

$$OCH_3$$
$$|$$
$$C=O$$
$$|$$
$$[(CH_2-CCl_2)-(CH_2-CH)] \text{ (VDC-MA)}$$

are presented. Note that both homopolymers PVDC and PMA have very similar T_g's, 2°C and 5°C, respectively. However, the VDC-MA copolymers exhibit significantly higher T_g's, which peak at about 40°C and 55°C for random and regularly alternating 50:50 VDC-MA copolymers. This unexpected behavior can be understood or at least rationalized based on the inherent conformational flexibilities of the copolymers and their constituent homopolymers. The conformational entropy S is employed to reflect the conformational flexibility, because the more flexible the polymer the more conformations it can adopt and the larger is its conformational entropy. To correlate the conformational entropies and T_g's of copolymers we assume that if the conformational entropy of a copolymer is greater/smaller than the compositionally weighted sum of the entropies of its constituent

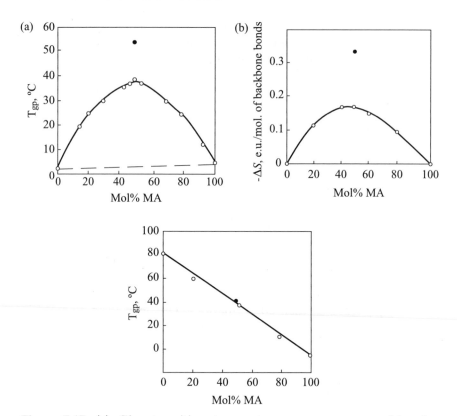

Figure 7.17. (a) Glass-transition temperature monomer composition behavior of VDC–MA copolymers (Hirooka and Kato, 1974): (○) random; (●) regularly alternating. (b) $\Delta S \equiv [X_{MA}S_{MA} + (1 - X_{MA})S_{VDC}] - S_{VDC-MA}$ calculated at 5°C as a function of monomer composition: (○) random; (●) regularly alternating (Tonelli, 1975). (c) Glass-transition temperature monomer composition behavior of VC–MA copolymers (Hirooka and Kato, 1974): (○) random; (●) regularly alternating.

homopolymers [i.e., $\Delta S = S_{VDC-MA} - X_{MA}S_{MA} - (1 - X_{MA})S_{VDC}$], then the copolymer T_g's will be lowered/raised in comparison to bulk additive behavior that would be manifested by the dashed curve in Figure 7.17(a). In other words, if ΔT_g is the deviation away from bulk additive T_g, then we expect $\Delta T_g \, \alpha - \Delta S$.

In Figure 7.17(b), $-\Delta S$ values calculated for VDC-MA are plotted against the copolymer compositon X_{MA}. ΔS values were calculated from the RIS model developed for VDC-MA copolymers (Tonelli,

TABLE 7.3. Conformational Entropies of VDC-MA Copolymer Chains of 500 Backbone Bonds

Mol% MA	Stereoregularity[a]	Sequence Distribution	S^b (eu/mol of Backbone Bonds)
0			0.754
20	A	Random	0.636[c]
40	A	Random	0.579[c]
50	A	Random	0.579[c]
50	A	Regularly alternating	0.412[c]
50	S	Regularly alternating	0.404
50	I	Regularly alternating	0.420
60	A	Random	0.591[c]
80	A	Random	0.641[c]
100	S		0.569
100	A		0.732
100	I		1.012

[a] S, syndiotactic; I, isotactic; a, atactic.

[b] Calculated at 5°C.

[c] Average of 10 Monte Carlo generated chains, where the mean deviation from the average entropy is ~1–2%.

1975) and are tabulated in Table 7.3, where it is clearly evident that the VDC-MA copolymer conformational entropies are significantly less than the weighted sum of VDC and MA homopolymer entropies. Thus we expect, as is observed (Hirooka and Kato, 1974), that the copolymer T_g's are raised above those expected for the bulk additive T_g's. In addition the entropy calculated for regularly alternating VDC-MA is smaller than that of the 50:50 random copolymer, so we also expect the former copolymer to have a T_g elevated above that of the latter as is also observed.

Table 7.4 and Figure 7.17(c) present the conformational entropies calculated for the related copolymer series vinyl chloride-methacrylate (VC-MA) and their observed T_g's, respectively. We can see from Table 7.4 that the copolymer and constituent homopolymer entropies are nearly coincident, so we expect, as is observed, very little deviation between the copolymer T_g's and those expected from the weighted addition of homopolymer T_g's. Thus, we are able to understand why adding MA/VDC to PVDC/PMA leads to increased T_g's, while adding MA/VC to PVC/PMA does not. These and several other exam-

TABLE 7.4. Conformational Entropies of VC-MA Copolymer Chains of 500 Backbone Bonds

Mol% MA	Stereoregularity[a]	Sequence Distribution	S^b (eu/mol of Backbone Bonds)
0	S		0.845
0	I		0.900
0	A		0.740[c]
50	S	Regularly alternating	0.768
50	I	Regularly alternating	0.854
50	A	Random	0.741[c]
100	S		0.560
100	I		1.015
100	A		0.725[c]

[a]S, syndiotactic; I, isotactic; a, atactic.

[b]Calculated at 40°C.

[c]Average of 10 Monte Carlo generated chains, where the mean deviation from the average entropy is ~1%.

ples (Tonelli, 1974a, 1975, 1977) have made apparent that the internal flexibility of copolymer chains as monitored by the RIS-calculated conformational entropies of their individual, isolated chains can explain their comonomer-sequence-dependent T_g's and provides another example of connecting the microstructures and bulk properties of polymers by utilizing knowledge of the microstructural sensitity of their conformational characteristics.

IMPACT STRENGTH OF POLYMERS

Polymers containing 1,4-phenylene linkages in their backbones can show remarkable impact strengths at temperatures well below their T_g's. For example, poly(2,6-dimethyl-1,4-phenylene oxide) (PPO), seen in Figure 7.18, shows a high-impact strength and resists brittle failure at room temperature, which is more than 175°C below its T_g. Because the phenyl groups in PPO are linked 1,4 in the backbone, its conformational characteristics can be developed (Tonelli, 1972, 1973) for the virtual bond representation of PPO seen in (b), where oxygen

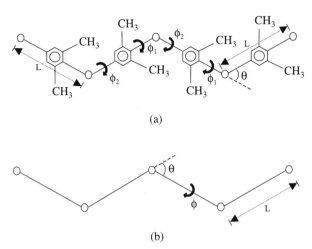

(a)

(b)

Figure 7.18. (a) A portion of the 2,6-dimethyl-1,4-phenylene oxide chain in the planar zigzag conformation, where $\phi_1 = \phi_2 = 0°$. All phenylene rings are coplanar in this reference rotational conformation, and ϕ_1 and ϕ_2 assume positive values for right-handed rotations. (b) A portion of the backbone of the 1,4-phenylene oxide chain planar zigzag conformation, where neighboring either oxygen atoms are connected by virtual bonds L. The virtual bond rotation angle is taken as $0°$ in this conformation and adopts positive values for right-handed rotations.

atoms are connected by virtual bonds $L = 5.5\,\text{Å}$. The net rotation angle around each virtual bond is a sum of two-rotation angles ϕ_2 and ϕ_1 about the real $O-C_1$ and C_4-O bonds. However, the rotation angles ϕ_1 and ϕ_2 on either side of a phenyl ring are independent of each other, because the intervening 1,4-phenylene ring separates the atoms to the right and left of the C_4 and C_1 atoms sufficiently to prevent their interaction. This is not the case for a pair of ϕ_1, ϕ_2 on either side of each oxygen.

The net result is that ϕ rotation around each virtual bond is a sum of two real rotation angles ϕ_2 and ϕ_1, whose energies are independent. From the conformational energy map for a pair of rotations ϕ_1, ϕ_2 on either side of the O's in PPO, we may obtain the probabilities for the net rotation ϕ about its virtual bonds as given in Table 7.5. Note that each value of ϕ is substantially populated and therefore accessible. This means that the PPO chain is nearly a truly freely rotating chain,

TABLE 7.5. Probabilities of Rotational States in Poly(2,6-dimethyl-1,4-phenylene oxide) ($\theta = 61°$)

ϕ	$f_\phi{}^a$	ϕ	$f_\phi{}^a$
0	0.036	180	0.036
20	0.063	200	0.063
40	0.060	220	0.060
60	0.073	240	0.730
80	0.036	260	0.036
100	0.036	280	0.036
120	0.073	300	0.073
140	0.060	320	0.060
160	0.063	340	0.063

[a] f_ϕ, fractional probability of rotational state ϕ.

where each value of the rotation angle ϕ is nearly equally likely. In other words, PPO is freely rotating in the dynamic, as well as the static, sense. Suppose the staggered *trans*, *gauche+*, and *gauche−* conformers in polyethylene had the same energies and were thus equally populated. Polyethylene would be freely rotating in the static sense, with a characteristic ratio $C_n = \langle r^2 \rangle_0 / nl^2 = 2.0$ (see Chapter 5). However, it would not be freely rotating in the dynamical sense, because only the three staggered conformations would be allowed, with energy barriers between them. In PPO there are no significant energy barriers between rotational states as evidenced from Table 7.5 (Tonelli, 1972, 1973).

If it is assumed that the impact strength of a polymer is related to the ability of its constituent chains to undergo rapid, reversible conformational transitions without bond rupture, then the nearly true free-rotation nature of PPO may be the source of its remarkable low temperature ($<T_g$) impact strength. We believe, therefore, that at least for polymers like PPO, such as polycarbonates and polysulfones, those bulk mechanical properties dependent upon high-frequency chain motions, like impact strength, are primarily governed by intramolecular barriers to backbone bond rotations. PPO, polycarbonate, and polysulfone possess no substantial rotation barriers, because of their symmetry and the separation of rotatable backbone bonds by 1,4-phenylenes, so they exhibit high impact strengths.

MODULUS OF AN ELASTOMERIC POLYMER NETWORK

Let us briefly analyze the thermodynamics of stretching a cross-linked polymer network. The work W required to stretch an individual chain between cross-links by an amount L by application of a force f is $f \cdot \Delta L$. From the first law of thermodynamics we know that the energy E of a system can only be changed by adding to or removing heat, Q or $-Q$, from the system or by performing work upon the system or having the system perform work, $-W$ or W (Castellan, 1983). As a consequence, $E = Q - W$. The second law of thermodynamics describes the connection between the change in entropy of a system ΔS produced by adding or removing heat; that is, $\Delta S = Q/T$. Combining the first and second laws leads to $\Delta E = T\Delta S - W = T\Delta S + f \cdot \Delta L$, or $f = (\Delta E - T\Delta S)/\Delta L$. We have mentioned in our observations of stretching a rubber band that upon stretching, both the width and thickness of the rubber band decrease, resulting in no overall change in sample volume V or $\Delta V \sim 0$. This implies that on average no change in the distances between polymer chains occurs on stretching and so any interchain contributions to the ΔE of stretching are negligible. As noted in Figure 7.7, the extension of a polymer chain between cross-links can generally be achieved by adjusting the conformations of only a very small number of the intervening bonds. Thus, the intrachain contributions to the ΔE of stretching are also normally expected to be negligible. We therefore expect the energy of a polymer network to be nearly independent of stretching, or $\Delta E/\Delta L \sim 0$. So the force required to stretch the polymer network reduces to $f = -(T\Delta S/\Delta L)$ and is entirely entropic in origin. We know $f > 0$ is required to stretch the polymer network, thus $\Delta S < 0$, or a reduction in entropy accompanies stretching, because the stretched network chains have fewer conformations available to them and hence a lower conformational entropy.

We have seen that when an amorphous, bulk polymer sample is cross-linked into a network, it exhibits remarkable elastic behavior when stretched above its T_g. Large reversible extensions are possible for small expenditures of force, with these elastic polymer networks having low moduli, because the retractive force is largely entropic in origin. However, what might be expected for a highly cross-linked network where the chains between cross-links are 10's and not 100's

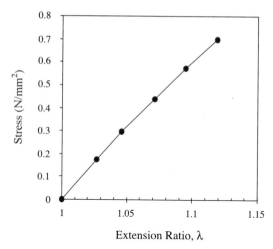

Figure 7.19. Stress–strain data for an end-linked PDMS network having a number average molecular weight between cross-links of 660 (9-mer) (Tonelli and Andrady, 1996).

of bonds in length? As the highly cross-linked network is stretched, a significant fraction of the bonds in the short chains between cross-links are required to change their conformations as they are extended. Thus we expect that the conformational entropy loss/bond to accompany extension would be much higher than that in a loosely cross-linked network where only a very small fraction of the network chain bonds need alter their conformations (see Figure 7.7). Because of the large fraction of bonds altering their conformations, we might also expect a significant energetic contribution to the retractive force exerted by a densely cross-linked polymer network.

Figure 7.19 presents the stress–strain behavior of a poly(dimethyl siloxane) (PDMS) network obtained by end-linking of the PDMS 9-mer ($n = 9$) with tetraethylorthosilicate [$(CH_3-CH_2-O-)_4-Si$]. The modulus for this dense network, whose short chains between cross-links possess only 16 conformable backbone bonds, is 2.2×10^6 N/m^2, which is considerably greater than that of a lightly cross-linked rubber band $\{[0.01 \text{ N}/\pi \, (0.5 \text{ mm})^2] = 10^4 \text{ N/m}^2\}$. Because the PDMS 9-mer contains only 16 conformable bonds between cross-links, it is possible to directly enumerate all 3^{16} of its RIS conformations and calculate the end-to-end distance **r**, the entropy S_{conf},

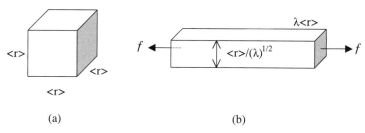

Figure 7.20. Three-chain network. (a) Unstrained state. (b) Simple extension.

and the energy E_{conf} of each conformation (Tonelli and Andrady, 1996). The internal force f required to stretch the 9-mer by Δr can be obtained, as discussed above, from $f = (\Delta E_{conf} - T\Delta S_{conf})/\Delta r$. To connect this force with that required to stretch a network composed of cross-linked 9mers, we adopt the three-chain network model (James and Guth, 1943) illustrated in Figure 7.20. Three 9-mer chains are arranged in a cube whose sides are $\langle\mathbf{r}\rangle$, $\mathbf{r}$ averaged over all 9-mer conformations, and each is assumed to be arranged with its end-to-end vector perpendicular to one of the three pairs of opposed cube sides. As we stretch the cube uniaxially, one of the three chains (the one along the stretch direction) is extended from $\mathbf{r} = \langle\mathbf{r}\rangle$ to $\lambda\langle\mathbf{r}\rangle$, while the other two are compressed from $\mathbf{r} = \langle\mathbf{r}\rangle$ to $\langle\mathbf{r}\rangle/(\lambda^{1/2})$, where $\lambda = L/L_0$, resulting in no over all change in sample volume.

In Figure 7.21, E_{conf} and S_{conf} calculated for the PDMS 9-mer from its RIS model (Flory et al., 1964) are presented as a function of the end-to-end length of the 9-mer. The force exerted by each chain is obtained from $(\Delta E_{conf} - T\Delta S_{conf})/\Delta r$, where $\mathbf{r}$ goes from $\langle\mathbf{r}\rangle$ to $\lambda\langle\mathbf{r}\rangle$ for the extended chain and from $\langle\mathbf{r}\rangle$ to $\langle\mathbf{r}\rangle/(\lambda^{1/2})$ for the two chains that are compressed. The total force $F_{tot} = f_{ext} + 2f_{comp}$ leads to the modulus $\sigma = F_{tot}/A_0(\lambda - 1)/(\lambda^2))$, where $A_0 = \langle\mathbf{r}\rangle^2$ is the cross-sectional area of the undeformed three-chain cube. Calculated moduli were in the range 10^6 to 10^7 N/m^2, which nicely bracket the experimental modulus of 2.2×10^6 N/m^2. Not only can we now understand why densely cross-linked polymer networks have much higher moduli than more typical loosely cross-linked networks, but once again we see a demonstration of how a bulk, solid-state polymer property (network elastic modulus) can be understood and calculated solely from knowledge of the conformational characteristics of its isolated, constituent, randomly coiling chains.

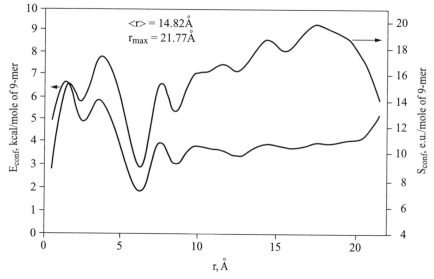

Figure 7.21. E_{conf} and S_{conf} calculated for a PDMS 9-mer (FCM-RIS model) as a function of the end-to-end 9-mer length r.

POLYMER MELTING

Finally let us revisit the topic of melting polymer crystals. We present a schematic representation of the melting of a polymer crystallite in Figure 7.22, where the melting process has been separated into two steps termed volume expansion and conformational disordering (Tonelli, 1970, 1974b). The former step is solely dependent on inter-actions between polymer chains, because of the attendant volume change and consequent increase in the separation between chains, while the latter occurs with no volume change and is intrachain in origin. Note that in the crystal and molten amorphous phases, the sample volumes are V_c and V_a, and the conformational energies and entropies of the individual polymer chains are E_c, E_a and S_c, S_a, respectively, while the expanded crystal is characterized by V_a, E_c, and S_c. Overall melting is characterized by $\Delta V = V_a - V_c$, $\Delta E = E_a - E_c$, and $\Delta S = S_a - S_c$, in addition to intermolecular contributions ΔE_{exp} and ΔS_{exp} due to ΔV. From thermodynamics we know that at $T = T_m$ the crystalline and amorphous phases in equilibrium must have the same free energy G, or $\Delta G_m = G_a - G_c = \Delta H_m -$

$$\Delta G_m = G_a - G_c = \Delta H_m - T_m \Delta S_m = 0$$

$$T_m = \Delta H_m / \Delta S_m$$

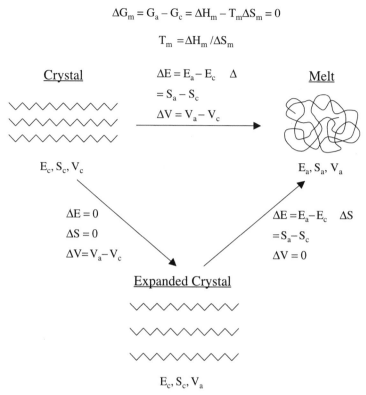

Figure 7.22. Decomposition of the melting of polymer crystals into two steps, volume expansion and conformational disordering, which are, respectively, predominantly inter- and intrachain processes (Tonelli, 1970, 1974b).

$T_m \Delta S_m = 0$, so that $T_m = \Delta H_m / \Delta S_m$. By examining the melting of polyethylene we may gauge the contributions made by the volume expansion and conformational disordering steps to the enthalpy and entropy of melting ΔH_m and ΔS_m.

Experimentally (Quinn and Mandelkern, 1958), $\Delta H_m = 960$ cal/mol of bonds, $T_m = 140°C$, and thus $\Delta S_m = 2.3$ cal/° K-mol (e.u.). If we ignore the very minor contribution to ΔH_m made by $\Delta V (P\Delta V)$, then $\Delta H_m = \Delta E_{exp} + \Delta E_{conf}$, which correspond to the volume expansion and conformational disordering steps, respectively. ΔE_{conf} is readily determined from our knowledge of the RIS model for randomly coiling polyethylene which revealed (see Chapter 5) that 40% of its bonds are in the *gauche* conformation having a 500 cal/mol higher energy than the *trans* conformation. As a result, ΔE_{conf} contributes 0.4×500 cal/mol $= 200$ cal/mol to ΔH_m, so that E_{exp} must be approximately $960 - 200 = 760$ cal/mol. For polyethylene the enthalpy of melting is dominated by the energy required to expand the sample from the crystal to the amorphous melt. From the conformational partition function Z for polyethylene, as obtained by matrix multiplication techniques and its RIS model, S_a can be calculated (Hill, 1962) for the conformationally disordered chains. If the fully extended, all-*trans* crystalline polyethylene chains are assumed to have no conformational entropy, then $\Delta S_{conf} = S_a = 1.8$ e.u., so $\Delta S_{exp} = \Delta S_m - \Delta S_{conf} = 2.3 - 1.8 = 0.5$ e.u. (Tonelli, 1970, 1974b). We see that for polyethylene the entropy of melting is dominated by the intrachain disordering of conformations.

As a consequence, we approximate $T_m = \Delta H_m / \Delta S_m$ by $T_m \sim \Delta E_{exp} / \Delta S_{conf}$, reflecting the observation, illustrated by the melting of polyethylene, that ΔH_m is dominated by changes in the interaction between chains accompanying the expansion of sample volume, while ΔS_m is dominated by the intramolecular process of conformational disordering of the polymer chains. $T_m \sim \Delta E_{exp} / \Delta S_{conf} = (760$ cal/mol$)/1.8$ e.u. $= 423$ K $= 150°$C, which approximates the $T_m = 140°$C observed for polyethylene (see Table 7.2). ΔE_{exp} would be expected to increase with the strength of interchain forces, such as dipolar, hydrogen-bonding, and van der Waals forces, while ΔS_{conf} should be smaller for those polymers that melt without assuming very many random-coil conformations. Consequently, T_m should be highest for polar, hydrogen-bonded polymers with conformationally stiff chains and should be lowest for nonpolar, flexible polymers.

To confirm the approach to the melting of polymers taken above, let us look at the results presented in Table 7.6 (Tonelli, 1974b). There the values of T_m, ΔS_m, ΔS_{exp}, and $(\Delta S_m)_V$, where $(\Delta S_m)_V = \Delta S_m - \Delta S_{exp}$, are presented for several crystalline polymers. In the final column of this table are the conformational entropies ΔS_{conf} calculated for these polymers from the appropriate RIS models. Except for *cis*-1,4-polyisoprene, there is a close correspondence between the experimentally determined $[(\Delta S_m)_V]$ and calculated (ΔS_{conf}) contributions made to the entropy of melting ΔS_m by the attendant conformational disordering of polymer chains. So, once again, we begin to understand an important characteristic *outside* response of bulk polymers, in this case the melting entropies and melting temperatures of polymers, by

TABLE 7.6. Entropies of Fusion (Tonelli, 1974b)

Polymer	T_m, °C	$\Delta S_m{}^a$	ΔS_{exp}	$(\Delta S_m)_V$	$\Delta S_{conf.}$
Polyethylene	140	2.29–2.34	0.46–0.52	1.77–1.84	1.76
Isotactic polypropylene	280	1.50	0.44–0.65	0.85–1.09	0.96
cis-1,4-Polyisoprene	28	0.87	0.45	0.43	1.34
trans-1,4-Polyisoprene	74	2.19	0.91	1.28	1.37
Polyoxymethylene	183	1.75	0.35	1.40	1.50
Polyoxyethylene	66	1.78	0.37	1.41	1.70
Polyethyleneterephthalate[b]	267	1.46	0.29	1.17	1.07
Polytetrafluoroethylene	327	1.97	0.52	1.45	1.60
cis-1,4-Polybutadiene	5	1.92	0.43	1.49	1.38
Polyethyleneadipate	65	1.48	0.38	1.10	1.04
Polyethylenesuberate	75	1.50	0.38	1.12	1.16
Polyethylenesebacate	83	1.54	0.38	1.16	1.24

[a] All entropies are given in eu/mol of backbone bonds.
[b] Benzene ring is treated as a single bond.

consideration of what polymer chains are doing individually on the *inside* (conformationally) during the process of melting.

DISCUSSION QUESTIONS

1. Unlike the case of most dilute polymer solutions, in amorphous, bulk polymer phases we consider the conformation of each chain to be unperturbed by excluded volume self-intersections. Why?

2. Why do the flows of small-molecule and low-molecular-weight polymer liquids show a relative insensitivity to the rate of shear characterizing their flow?

3. What dominant factor makes prediction of the glass-transition temperature of a polymer difficult?

4. Suppose a polymer is cross-linked while being deformed, say while being extruded through a narrow orifice as found in a fiber spinnerette. We know that the extruded polymer liquid will puddle up after falling on a flat surface if it is not cross-linked, but will the polymer that is cross-linked during extrusion also puddle up? Why or why not?

5. A polymer is cross-linked in solution and in a second experiment it is cross-linked in the bulk, using the same cross-linking agent in both instances. Each experiment is conducted in a manner that yields polymer networks possessing the same cross-link density. Which network (solution or bulk cross-liked) would be expected to have the larger modulus and why?

6. Suggest explanations for the general observation that when polymers crystallize they produce semicrystalline materials, with both amorphous and crystalline domains coexisting in the same sample.

7. Begin consideration of a bulk polymer sample that is composed of crystallizable chains that is above its T_m and then is gradually cooled first below T_m and then below T_g. Describe physical behaviors expected from this sample when observed at $T > T_m$, $T_g < T < T_m$ and $T < T_g$, and give reasons for your expectations.

*8. Jello or gelatin is an example of a highly swollen polymer network consisting of collagen protein chains and water. As we all know, it is an elastic material (it jiggles) that can be converted to a liquid by increasing the temperature and reversibly returns to the gelled state upon cooling, characteristic of polymer materials called thermoreversible gels. Speculate on the structure of gelatin with emphasis on those features that might be expected to result in the behavior outlined above. Also compare and contrast the behaviors and structures of gelatin and slime, both of which are at least partially elastic.

*9. Using the first and second laws of thermodynamics, analyze the stretching of an elastic polymer network. Remembering that this is a process characterized by constant volume and energy, derive an expression for the retractive force and its temperature dependence.

*10. Based on our experience with the conformations, motions, and interactions of polymer chains as they randomly coil in solution and in the melt, suggest reasons why the chain-folded lamellar morphology predominates in both solution and melt-grown crystallization of polymers.

*11. We know that the randomly coiling chains in a polyethylene melt are characterized by dimensions ($\langle r^2 \rangle_0$) that correspond to $C_n = \langle r^2 \rangle_0 / nl^2 = 6.8$. If we are spinning polyethylene fibers, what would be the maximum draw ratio expected for a sample with a molecular weight of 1,000,000?

*12. In our discussion of the possible connection between the conformational entropies and T_g's of copolymers and their dependence on comonomer sequence distribution, it was suggested that if the conformational entropy of a copolymer is larger than the weighted sum of constituent homopolymer conformational entropies, its T_g will be lower than that of the weighted average of homopolymer T_g's. What implicit assumption serves as the basis for this suggestion, and why do you believe it to be/or not to be reasonable?

*13. It was argued that the nearly truly, or dynamically, freely rotating character of polymers like PPO lies at the heart of their high impact strengths evidenced well below their T_g's. Beginning from this frame of reference, comment on why and/or how they also have high T_g's?

*14. Aliphatic polyamides and polyesters made from identical diacids and analogous diamines and diols with the same numbers of methylene groups show melting temperatures that differ by more than 150°C, with the polyamides always melting at the elevated temperatures. On the other hand, the analogous polyamides and polyesters melt with very similar ΔH_m's, and the random coiling of their molten chains is governed by very similar RIS models. Based on this information, suggest a rational explanation for the large difference between the T_m's of aliphatic polyamides and polyesters.

REFERENCES

Abe, Y., and Flory, P. J. (1970), *J. Chem Phys.*, **52**, 2814.

Bloomfield, G. F. (1946), *J. Polym. Sci.*, **1**, 312.

Bovey, F. A. (1979), *Macromolecules: An Introduction to Polymer Science*, F. A. Bovey and F. H. Winslow, Eds., Academic Press, New York, Chapter 5.

Bowden, F. C., Price, N. W., Bernal, J. D., Fankuchen, I. (1936), *Nature*, **138**, 1051.

Bueche, F. (1962), *Physical Properties of Polymers*, Interscience, New York.

Casassa. E. Z., Sarquis, A. M., and Van Dyke, C. H. (1986), *J. Chem. Ed.*, **63**, 57.

Castellan, G. W. (1983), *Physical Chemistry*, 3 ed., Benjamin-Cummings, Menlo Park, CA.

Chiang, R., and Flory, P. J. (1961), *J. Am. Chem. Soc.*, **83**, 2857.

deGennes, P. (1971), *J. Chem. Phys.*, **55**, 572.

deGennes, P. (1979), *Scaling Concepts in Polymer Physics*, Cornell University Press, Ithaca, NY.

deGennes, P. (1983), *Physics Today*, **36**, 33.

Doi, M., and Edwards, S. F. (1978), *J. Chem. Soc. Faraday Trans. II*, **74**, 1789, 1802, 1818.

Edwards, S. F. (1967), *Proc. Phys. Soc.*, **92**, 9.

Farmer, E. H., and Shipley, F. W. (1946), *J. Polym. Sci.*, **1**, 293.

Ferry, J. D. (1980), *Viscoelastic Properties of Polymers*, 3 ed., John Wiley and Sons, New York.

Fetters, L., J., Lohse, D. J., Richter, D., Witten, T. A., and Zirkel, A. (1994), *Macromolecules*, **27**, 4639.

Flory, P. J. (1953), *Principles of Polymer Chemistry*, Cornell University Press, Ithaca, NY.

Flory, P. J. (1962), *J. Am. Chem. Soc.*, **84**, 2857.

Flory, P. J., Crescenzi, V., and Mark, J. E. (1964), *J. Am. Chem. Soc.*, **86**, 146.

Gans, C. (1920), *Scientific American*, **82**.

Gordon and Macnab (1953).

Graessley, W. W. (1974), *Adv. Polym. Sci.*, **16**, 1.

Graessley, W. W. (1982), *Adv. Polym. Sci.*, **47**, 67.

Hill, T. L. (1962), *Introduction to Statistical Thermodynamics*, Addison Wesley, Reading, MA.

Hirooka, M., and Kato, T. (1974), *J. Polym. Sci. Polym. Lett. Ed.*, **12**, 31.

James, H. M., and Guth, E. (1943), *J. Chem. Phys.*, **11**, 455.

Keller, A. (1957), *Philos. Mag.*, **2**, 1171.

Lin, Y.-H., and Juang, J.-H. (1999), *Macromolecules*, **32**, 181.

Mandelkern, L. (1972), "An Introduction to Macromolecules", Springer-Verlag, New York.

Marrucci, G., and deCindo, G. (1980), *Rheol. Acta*, **19**, 68.

Marrucci, G., and Hermans, J. J. (1980), *Macromolecules*, **13**, 380.

Marrucci, G. (1981), *Macromolecules*, **14**, 434.

Naylor, R. F. (1946), *J. Polym. Sci.*, **1**, 305.

Perkins, T. T., Smith, D. E., and B. Chu (1994), *Science*, **264**, 819.

Quinn, F. A., Jr., and Mandelkern, L. (1958), *J. Am. Chem. Soc.*, **80**, 3187.

Rottach, D. R., Tillman, P. A., McCoy, J. D., Plimpton, S. J., and Curro, J. G. (1999), *J. Chem. Phys.*, **111**, 9822.

Srinivasaro, M. (1995), *Int. J. Mod. Phys. B*, **9**, 2518.

Starkweather, H. W., Jr., Zoller, P., Jones, G. A., and Vega, A. J. (1982), *J. Polym. Sci., Polym. Phys. Ed.*, **20**, 751.

Storks, K. H. (1938), *J. Am. Chem. Soc.*, **60**, 1753.

Tonelli, A. E. (1970), *J. Chem. Phys.*, **52**, 4749.

Tonelli, A. E. (1972), *Macromolecules*, **5**, 558.

Tonelli, A. E. (1973), *Macromolecules*, **6**, 503.

Tonelli, A. E. (1974a), *Macromolecules*, **7**, 632.

Tonelli, A. E. (1974b), *Analytical Calorimetry*, Vol. 3, R. S. Porter and J. F. Johnson, Eds., Plenum, New York, p. 89.

Tonelli, A. E. (1975), *Macromolecules*, **8**, 544.

Tonelli, A. E. (1977), *Macromolecules*, **10**, 716.

Tonelli, A. E., and Andrady, A. L. (1996), *Comp. Theor. Polym. Sci.*, **6**, 103.

Treloar, L. R. G. (1975), *The Physics of Rubber Elasticity*, 3 ed., Clarendon Press, Oxford, United Kingdom.

CHAPTER 8

NATURALLY OCCURRING BIOPOLYMERS

POLYSACCHARIDES

The most abundant naturally occurring biopolymers belong to the class called polysaccharides (or polysugars). In Figure 8.1 the linking of sugars to form polysaccharides is indicated, where both 1,4-α and 1,4-β linkages are illustrated. Cellulose, the most abundant organic compound on Earth, is the polysaccharide obtained through the 1,4-β linking of glucose sugar units, as shown in Figure 8.2, where amylose starch, consisting of 1,4-α linked glucose sugar units, is also illustrated. On the surface it might seem that the difference in structures between cellulose and starch is relatively minor (1,4-β versus 1,4-α linkage of glucose); however, these polysaccharides exhibit widely different chemical, physical, and biological properties.

Cellulose is a generally insoluble, infusible crystalline polymer that serves a structural role in nature. Wood, cotton, and other plant cell walls and fibers are made from cellulose. Native cellulose is usually about 70% crystalline and ignites before it melts above 280°C (Daniel, 1990). As a consequence, very little cellulose consumed by humans is in a form other than those found naturally in foods, fuels, lumber, and fibers. Purifed cellulose (i.e., cellulose isolated from other con-

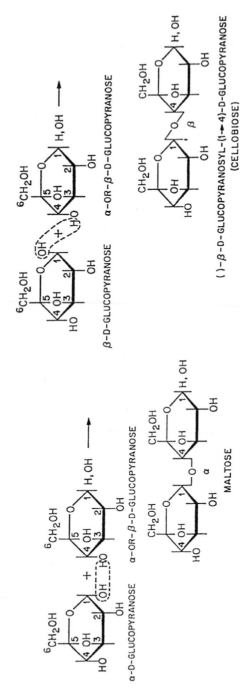

Figure 8.1. Illustration of the formation of $1 \rightarrow 4$ α- and β-glycosidic bonds joining two sugars in disaccharides. [Adapted from MacGregor and Greenwood (1980).]

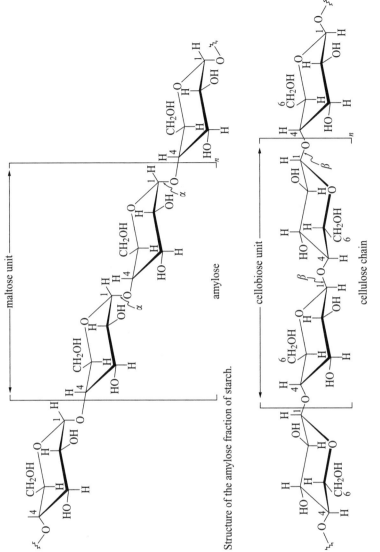

Structure of the amylose fraction of starch.

Partial structure of a cellulose molecule showing the β linkages of each glucose unit.

Figure 8.2. Starch(amylose) and cellulose structures.

sitituents found in naturally occurring sources), is predominantly utilized in the production of paper, paperboard, and textiles.

The three hydroxyl groups on each glucose unit and the conformational restrictions placed on the cellulose chain by their 1,4-β linkages (see below) result in efficient packing and strong interactions between cellulose chains. It is thus likely (Tonelli, 1974b) that $T_m \sim \Delta E_{exp}/\Delta S_{conf}$ is very high, because ΔE_{exp} is large due to extensive interchain hydrogen-bonding between hydroxyl groups and ΔS_{conf} is small due to the limited number of conformations available to a cellulose chain. The net result is a melting temperature that exceeds the decomposition temperature and also suggests a reason why cellulose is very difficult to dissolve. Though difficult to dissolve, cellulose may be significantly swollen, which reduces crystallinity and increases accessibility, a phenomenon important to its processing into pulp, paper, and textiles and to its physical properties. For example, fabric made from cellulose fibers is characterized by its ability to absorb and wick away, or remove by capillary action, any moisture that it contacts. For this reason, cotton fabrics are often preferred for wear in tropical climates.

In addition to natural forms, such as wood and cotton fibers, cellulose can be chemically modified, principally through reaction of its hydroxyl groups, to form polysaccharides with a wide range of properties and applications. For example, cellulose nitrate (TNT) and cellulose acetate obtained by the nitration and acetylation of cellulose with HNO_3 and CH_3COOH, find use, respectively, in explosives and as a soluble cellulose derivative from which both fibers and films are produced. Cellulose fibers, other than those obtained directly from cotton and other plants, can be spun from solution (Turbak, 1990) either directly from N-methylmorpholine N-oxide or after achieving solubility by derivativation in the viscose process. In the latter process, various natural cellulose sources, such as wood pulp, are derivatized to induce solubility, spun from the resulting solution into fibers, and "de-derivatized" to regenerate cellulose in fiber form, which is generally called rayon. Due to the large amounts of water required and the impact of potential pollutants generated by the viscose-rayon process, it is no longer used in America to produce cellulose fiber.

Starch is the principal food-reserve polysaccharide in plants, and it also serves as the main source of carbohydrates in the diets of animals, including humans, in the form of the cereal grains corn, rice, and wheat and from potatoes. In addition to the linear, 1,4-α linked polysaccharide amylose (see Figure 8.2), starch also contains the branched polysaccharide amylopectin, with some 1,6-α linked glucose rings. Although amylose and cellulose have identical structures, aside from their modes of glycosidic linkage, amylose is inherently more flexible than cellulose (Brant and Christ, 1990) and, unlike cellulose, shows limited aqueous solubility. In fact, the characteristic ratios $C_n = \langle \mathbf{r}^2 \rangle_0 / nl^2$, where l is the length of the virtual bond connecting glycosidic oxygens, measured (Sarko, 1976; Okano and Sarko, 1984) for amylose and cellulose (soluble ester or ether derivatives) are 5 and 36, respectively, confirming the greater flexibility of amylose.

The dimensions calculated for amylose and cellulose from conformational energy maps $E(\phi, \psi)$, where ϕ, ψ are the rotational angles about the C_1–O and O–C_4 glycosidic bonds and glucose rings are rigidly held in the chair conformation, reproduce (Brant and Christ, 1990) the observed dimensions. A more graphic impression of the increased conformational flexibilty of amylose can be obtained from Figure 8.3, where example trajectories of amylose and cellulose chains generated from their conformational energy maps are presented. Here it is clearly evident that the amylose chain is much more randomly coiling than cellulose.

An additional important difference between amylose or starch and cellulose, which is also likely due to their different modes of glucose ring connectivity, is their comparative susceptibility to digestion by specific enzymes. (Enzymes are proteins, which we discuss below, that can catalyze specific chemical reactions.) For example, most animals are able to digest starch and use it as a carbohydrate source, but at the same time cannot digest cellulose, which passes intact through their bodies. We have all heard about the purported health benefits of a diet high in fiber. Well, this dietary fiber is cellulose, which humans cannot digest. On the other hand, termites possess an enzyme that enables them to very efficiently digest cellulose and convert it into a food source for carbohydrates, giving a whole new meaning to the phrase "eating me out of house and home."

The textile industry takes advantage of the specificity of enzymatic digestion of the polysaccharides amylose or starch and cellulose. Cotton, cellulosic yarns are often coated with a thin layer of starch in a process called *sizing* (Hackin, 1980). A sized yarn is smoother and less hairy and, as a consequence, may be woven or knitted into fabric

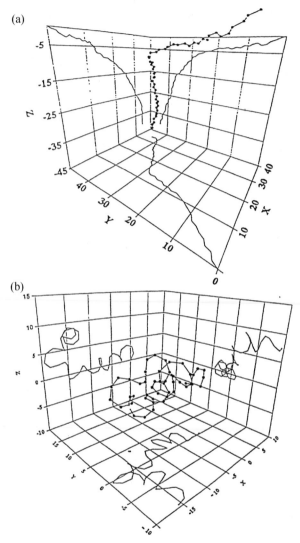

Figure 8.3. Snapshots of typical cellulosic (a) and amylosic (b) chain trajectories taken from Monte Carlo samples based on their rigid glucose ring conformational energy maps (Brant and Christ, 1990). Filled circles represent glycosidic oxygens, linked by virtual bonds L spanning the sugar residues (not shown), permit the tracing of the instantaneous chain trajectories in a coordinate system rigidly fixed to one chain end. Visualization is assisted by projections of the chains onto three mutually orthogonal planes.

at a faster rate and with reduced breakage compared to its unsized counterpart. However, after construction of the fabric, the size must be removed in order to recapture the beneficial qualities of the cellulosic cotton. Of course, scouring with a strongly basic aqueous solution will remove the starch, but not without degradative hydrolysis of both the starch and the cellulose. Instead, if an amylase enzyme is used to treat the fabric sized with starch, the starch is removed by digestion without any deleterious effect on the underlying cellulose fibers in the cotton fabric.

Before bringing our discussion of polysaccharides to a close, we mention two cellulose derivatives, chitin, ubiquitously found in nature, and chitosan a man-made derivative of chitin. Chitin is identical to cellulose save for the substitution of the hydroxyl group on the 2-carbon with an N-acetyl group ($-NH-\overset{\overset{O}{\|}}{C}-CH_3$). Chitin is found in the cell walls of mushrooms and in crustaceans and most insects, where it forms the protective outer layer of their shells. After cellulose, chitin is the most abundant organic material on Earth. Chitin also shares with cellulose an intractable nature, but can be readily converted into a soluble, more processable form called chitosan (Rathke and Hudson, 1994). Treatment of chitin with NaOH solution converts the N-acetyl groups into amine ($-NH_2$) groups and confers on the resultant chitosan solubility in several benign solvents including both formic and acetic acids. This solubility has led to the production of chitosan fibers and films, which may become commercially significant, because chitosan has been demonstrated to be biodegradable/bioabsorbable, to accelerate wound healing and the germination of seeds, and is able to remove heavy metals by chelation from contaminated process streams (Tokura et al., 1987; Hirano, et al., 1989; Rorrer, et al., 1993; Rathke and Hudson, 1994).

POLYPEPTIDES AND PROTEINS

Polypeptides and proteins are polyamides made from amino acids with the following structures: $NH_2-\overset{\overset{R}{|}}{\underset{\underset{H}{|}}{C}}-COOH$. Consequently, these

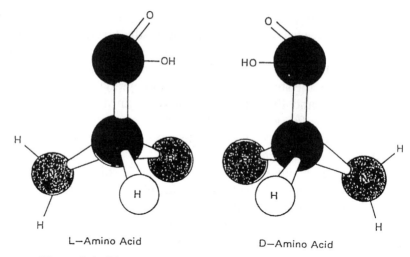

L—Amino Acid   D—Amino Acid

Figure 8.4. The two stereoisomeric forms of an amino acid.

polymers consist of chains of amide bonds each separated by a single, substituted, tetrahedral carbon atom. Unless $R = H$, which is the case for the amino acid glycine, amino acids are chiral molecules, because four distinct atoms or groups are bonded to the C^α making it an asymmetric center with a non-superimposible mirror image and hence optically active. Figure 8.4 illustrates the two sterioisomeric forms of an amino acid, and Table 8.1 presents the structures for the 20 amino acids found in proteins and used to make polypeptides. Polypeptides generally are made from a single amino acid, while proteins may contain one or more residues from all 20 amino acids. All proteins found in nature are made from L-amino acids, though several small biologically active polypeptides that function as hormones, toxins, and regulating factors and contain more than a single amino acid are known to have some D-residues as well (Tonelli, 1986).

Because proteins are built up from 20 different amino acids, a truly astronomical number of unique proteins are possible. For example, a protein containing 100 amino acid residues has 20^{100} or $\sim 10^{130}$ distinguishible variants. Thus, it is not so surprising to find that proteins serve in such a wide range of diverse roles in the biological world (Dickerson and Geis, 1969; Bell and Bell, 1988). They form the structural materials that hold living organisms together, and within living organisms they transport molecular and ionic species, control

TABLE 8.1. Amino Acids Found in Proteins

Amino Acid	Chemical Structure	Amino Acid	Chemical Structure
Glycine (Gly)	$\overset{H}{\underset{NH_2}{H-C-\overset{O}{\underset{}{C}}-OH}}$	Glutamic (Glu)	$\overset{O}{\underset{HO}{C}}-CH_2-CH_2-\overset{H}{\underset{NH_2}{C}}-\overset{O}{\underset{}{C}}-OH$
Alanine (Ala)	$CH_3-\overset{H}{\underset{NH_2}{C}}-\overset{O}{\underset{}{C}}-OH$	Lysine (Lys)	$NH_3-CH_2-CH_2-CH_2-CH_2-\overset{H}{\underset{NH_2}{C}}-\overset{O}{\underset{}{C}}-OH$
Valine (Val)	$\underset{CH_3}{\overset{CH_3}{>}}CH-\overset{H}{\underset{NH_2}{C}}-\overset{O}{\underset{}{C}}-OH$	Arginine (Arg)	$NH_2-\overset{\underset{\parallel}{C}}{\underset{NH}{}}-NH-CH_2-CH_2-CH_2-\overset{H}{\underset{NH_2}{C}}-\overset{O}{\underset{}{C}}-OH$
Leucine (Leu)	$\underset{CH_3}{\overset{CH_3}{>}}CH-CH_2-\overset{H}{\underset{NH_2}{C}}-\overset{O}{\underset{}{C}}-OH$	Histidine (His)	$HC=C-CH_2-\overset{H}{\underset{NH_2}{C}}-\overset{O}{\underset{}{C}}-OH$, $\underset{\underset{H}{C-}}{N \; NH}$
Icoleucine (Ileu)	$CH_3-CH_2-CH-\overset{H}{\underset{CH_3\,NH_2}{C}}-\overset{O}{\underset{}{C}}-OH$	Methionine (Met)	$CH_3-S-CH_2-CH_2-\overset{H}{\underset{NH_2}{C}}-\overset{O}{\underset{}{C}}-OH$
Serine (Ser)	$HO-CH_2-\overset{H}{\underset{NH_2}{C}}-\overset{O}{\underset{}{C}}-OH$		

continued

TABLE 8.1. (continued)

Amino Acid	Chemical Structure	Amino Acid	Chemical Structure
Threonine (Thr)	$CH_3-\overset{\overset{H}{\mid}}{C}-\overset{\overset{H}{\mid}}{C}-\overset{\overset{O}{\parallel}}{C}-OH$ with OH and NH_2	Cysteine (CySH)	$HS-CH_2-\overset{\overset{H}{\mid}}{C}-\overset{\overset{O}{\parallel}}{C}-OH$ with NH_2
Tyrosine (Tyr)	$HO-\langle\bigcirc\rangle-CH_2-\overset{\overset{H}{\mid}}{C}-\overset{\overset{O}{\parallel}}{C}-OH$ with NH_2	Cystine (Cys)	$S-CH_2-\overset{\overset{H}{\mid}}{C}-\overset{\overset{O}{\parallel}}{C}-OH$ with NH_2 ; $S-CH_2-\overset{\overset{H}{\mid}}{C}-\overset{\overset{O}{\parallel}}{C}-OH$ with NH_2
Phenylalanine (Phe)	$\bigcirc-CH_2-\overset{\overset{H}{\mid}}{C}-\overset{\overset{O}{\parallel}}{C}-OH$ with NH_2	Proline (Pro)	CH_2-CH_2 ; $CH_2 \quad CH-\overset{\overset{O}{\parallel}}{C}-OH$; $N-H$
Tryptophan (Try)	$\bigcirc-\overset{}{C}-CH_2-\overset{\overset{H}{\mid}}{C}-\overset{\overset{O}{\parallel}}{C}-OH$ with NH_2 ; $=CH$; $N-H$	Hydroxyproline (Hypro)	$HO-CH-CH_2$; $CH_2 \quad CH-\overset{\overset{O}{\parallel}}{C}-OH$; $N-H$
Aspartic (Asp)	$\overset{\overset{O}{\parallel}}{C}-CH_2-\overset{\overset{H}{\mid}}{C}-\overset{\overset{O}{\parallel}}{C}-OH$ with HO and NH_2		

and catalyze virtually all metabolic processes, and combat bacterial and viral diseases in the form of antibodies. The functional versatility of proteins stems not solely from their great chemical diversity, as attested to by the wide range of side groups possessed by the 20 amino acid building blocks presented in Table 8.1, but because polypeptide and protein chains also possess a pronounced diversity of potential conformations as well (Flory, 1969; Grassberger et al., 1998).

This may at first seem surprising, given that polypeptide and protein backbones are nothing more than a chain of amide bonds connected by single carbon atoms. Amide bonds possess double bond character, are not readily rotatable, and therefore are generally restricted to the planar, *trans* conformation. As a result, only two of the three bonds in each amino acid residue are rotatable and may change conformations; that is, the $-\underset{\underset{H}{|}}{N}-C-$ and $-C-\underset{\overset{||}{O}}{C}-$ bonds are conformationally flexible, while the $-\underset{\underset{H}{|}}{N}-\overset{\overset{O}{||}}{C}-$ amide bond is rigid. Figure 8.5 presents a portion of a poly-L-peptide chain, where the geometry, bond lengths and valence angles, and rotation angles are defined and indicated. Though the amide bonds are prevented from rotating away from their planar, *trans* conformation, which reduces the potential conformational diversity of polypeptides and proteins, at the same time the separation of atoms on opposites sides of the amide bonds generally renders the rotations (ϕ_i, ψ_i) within a residue and their energies independent of the conformations of neighboring residues, such as (ϕ_{i-1}, ψ_{i-1}) and (ϕ_{i+1}, ψ_{i+1}).

An indication of the potential conformational diversity of polypeptides and proteins can be obtained from the conformational energy maps $[E(\phi, \psi)$ versus $(\phi, \psi)]$ of their amino acid residues. Two such maps are presented in Figure 8.6 for the glycine and L-alanine amino acid residues, where the interactions between all atoms spanned by and including C_{i-1}^{α} and C_{i+1}^{α} were considered in the calculation of $E(\phi_i, \psi_i)$ (see Figure 8.5). As can be readily seen in both conformational energy maps, neither the glycine or the L-alanine residue is exclusively limited to the staggered *trans*, *gauche*+, and *gauche*− bond rotation conformers as most synthetic polymers are (see Chapter 5). This is a consequence of the low inherent rotational

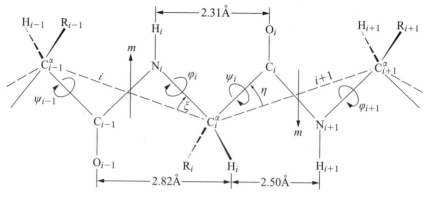

Bond lengths, Å		Bond angle and its upplement (θ), deg[c]	
$C^\alpha–C$,	1.53	$\angle C^\alpha CN$,	66
$C–N$,	1.32	$\angle CNC^\alpha$,	57
$N–C^\alpha$,	1.47	$\angle NC^\alpha C$,	70
$C=O$,	1.24		
$N–H$,	1.00		
$C^\alpha–C^{\beta,b}$,	1.54		
$C^\alpha–H^{\alpha}$,	1.07[d]		

[a] $\eta = 22.2°$; $\xi = 13.2°$; l_u = length of virtual bond = 3.80 Å.

[b] C^β denotes the first carbon of the substituent R, assuming R to be $-CH_2-R'$.

[c] Angles for bonds to pendant atoms (O, H, R) may be assumed, in legitimate approximation, to subdivide equally the ranges available to them.

Figure 8.5. Geometrical representation of a portion of a poly-L-peptide chain in the all-*trans* ($\phi = \psi = 0°$), planar conformation. Bond lengths and valence angles are listed below, and the amide dipole moments and virtual bonds connecting neighboring C^α's are indicated by *m*'s and dashed lines, respectively. [Adapted from Flory (1969).]

barriers about the $-\underset{\underset{\text{H}}{|}}{\text{N}}-\overset{\overset{\text{O}}{\|}}{\text{C}}-$ and $-C-C-$ bonds, which are ~1 kcal/mole (Brant and Flory, 1965) and therefore much lower than the 3.5-kcal/mole barrier to rotation about the $-C-C-$ bond in butane (see Figure 5.2) and most all carbon backbone polymers.

When polypeptides are dissolved and proteins are dissolved and denatured, they adopt randomly coiling conformations that are

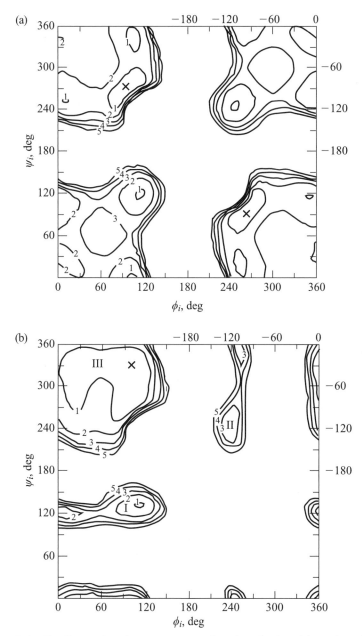

Figure 8.6. Conformational energy maps for the glycine (a) and L-alanine (b) residues, where the energy $E(\phi, \psi)$ contours are drawn at 1-kcal/mol intervals (Brant et al., 1967).

consistent with the conformational energy maps of their constituent amino acids, such as those presented for Gly and L-Ala in Figure 8.6 (also see Figures 8.21 and 8.22 below). For example, the characteristic ratios $C_n = \langle r^2 \rangle_0 / nL^2$, where n is the number of amino acid residues and L is the length of the virtual bond connecting C^α carbons in adjacent residues (see Figure 8.5), calculated for polyglycine, poly-L-alanine, and the denatured silk protein are 2.2, 9.3, and 2.4, respectively (Brant and Flory, 1965; Mathur et al., 1997). The unperturbed dimensions measured for polypeptides like poly-L-alanine, with R = –CH$_2$–X, are $C_n = 9.0 \pm 0.5$; for denatured silk protein, $C_n = 2.4$ (Jackson and O'Brien, 1995; Mathur et al., 1997). Not only are polypeptides flexible random coils in solution, but structural or fibrous, and even globular, proteins when denatured transform from either extended, regularly repeating or compact, globular conformations characterized by IQ ~ 1 to randomly coiling polymer chains with IQ ~ 100.

The fact that polypeptides and proteins are potentially more conformationally diverse than most synthetic polymers is not important in terms of their disordered, randomly coiling behavior, but instead is critical to the conformations they may adopt in their ordered states, because proteins function not as disordered random coils, but as conformationally ordered macromolecules. Proteins either adopt regularly repeating conformations with extended sizes and shapes when their functions are structural, or they adopt regular though not repeating conformations and globular overall shapes when they function as catalysts, transport agents, regulators, or antibodies (Dickerson and Geis, 1969; Bell and Bell, 1988).

Structural or fibrous proteins adopt regular, repeating, extended conformations with each of their amino acid residues in the same conformation. Figure 8.7 depicts fully extended protein chains, with each residue in the all-*trans* conformation, in a hydrogen-bonded sheet structure as found in silk proteins. This is very reminiscent of the crystalline structure of aliphatic polyamides or nylons. A comparison of the Pauling α-helical conformation for polypeptides and proteins and the 3_1-helix found in the crystals of several isotactic vinyl polymers is made in Figure 8.8. Note the intrachain hydogen bonding in the α-helix, in contrast to the interchain hydrogen bonding in the β-sheet, which stabilizes this conformation. Three α-helical protein chains may be wound into a protein rope, and this constitutes the

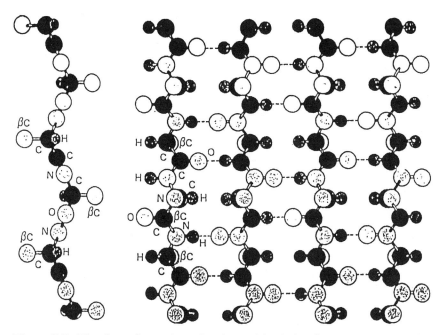

Figure 8.7. The β conformation of polypeptide chains (Marsh et al., 1955).

α-keratin protofibril found in hair and wool, as indicated in Figure 8.9. Also seen in Figure 8.9(b) is the cross section of an α-keratin fiber as observed with an electron microscope. Each small circle represents the three-stranded, α-helical protofibril seen in (a), which are arranged further into an 11-stranded cable or microfibril embedded in an amorphous, disordered protein matrix. A human hair contains hundreds of the α-keratin micofibrils joined together.

Figure 8.10 includes a drawing of the left-handed helical structure adopted by the homo-polypeptide poly-L-proline when crystallized (Cowan and McGavin 1955; Sasisekharan, 1959; Sarko, 1976). The left-handed structure has 3 residues/turn in contrast with the right-handed. α-helix, which contains 3.5 residues/turn (see Figure 8.8). The protein collagen is the fibrous component in cartilage, ligaments, and tendons, and it serves as the scaffold upon which the mineral hydroxyapatite is deposited and crystallized to form bone (Matheja and Degens, 1971). Every third residue in collagen is glycine, and 20–30% of the residues are either proline or hydroxy proline. Gly-X-Pro or Hypro and Gly-Pro-Hypro sequences predominate and apparently

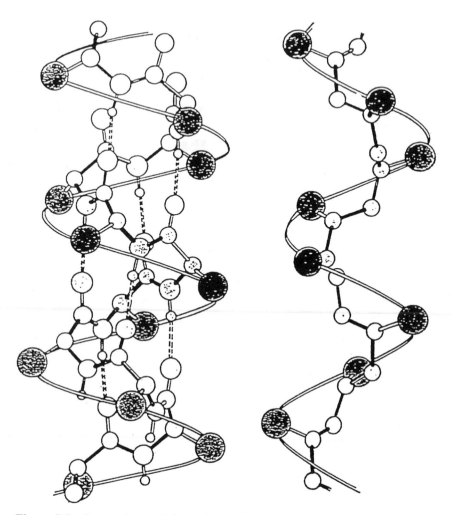

Figure 8.8. Comparison of the polypeptide α-helix (Pauling, 1960) and the 3_1-helix of crystalline isotactic vinyl polymers (Natta and Corradini, 1960).

result in collagen adopting the left-handed poly-L-proline helical conformation. Three left-handed collagen helices entwine to form tropocollagen as seen in parts (b) and (c) of Figure 8.10. The collagen fibrils seen in part (d) are composed of tropocollagen ropes that are aligned and regularly displaced from one another by one-fourth of their length, and this packing arrangement results in the characteristic banding pattern observed in collagen fibrils (Rice, 1960).

(a)

(b)

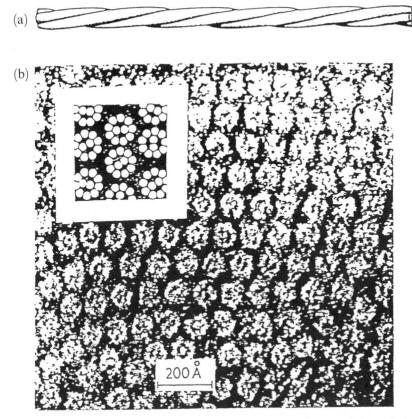

Figure 8.9. (a) Alpha-keratin protofibril formed by the entwining of three α-helical keratin protein chains (Pauling and Corey, 1953). (b) Electron micrograph of the cross section of an α-keratin fiber (Filshie and Rogers, 1961). Note that the inset is a schematic representation of the electron micrograph results.

The fibrous or structural proteins adopt regular repeating conformations and organize and pack in manners similar to that in synthetic polymer fibers, and both classes of macromolecular material show similar physical properties (Mandelkern, 1972). Globular proteins, on the other hand, adopt conformations and overall sizes and shapes that are not encountered in synthetic polymers (Flory, 1969; Mandelkern, 1972). Globular proteins adopt compact conformations with very little void volume within their spatial domains. In other words, even though they are not highly extended, their IQ's approach

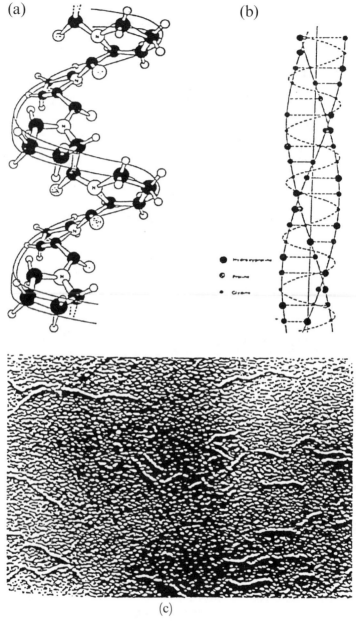

Figure 8.10. (a) Left-handed helical conformation of poly-L-proline. (b) Triple-stranded tropocollagen (Holum, 1994). (c) Electron micrographs of calf skin tropocollagen and (d) collagen fibrils (Rice, 1960).

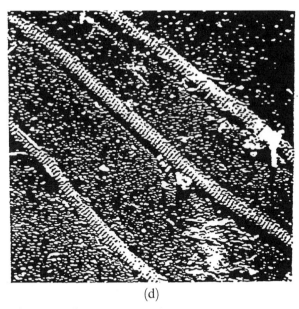

(d)

Figure 8.10. (continued)

unity. Thus, unlike randomly coiling polymers characterized by high IQ's, the external influence of tightly coiled globular proteins is essentially limited to the volume they physically occupy, which limits the cooperative nature of their behaviors. The tight packing of atoms in globular proteins is clearly demonstrated in Figure 8.11, where a space-filling model of the cell-digesting enzyme lysozyme is presented. This model is based on analysis of diffraction data observed (Dickerson and Geis, 1969) from the scattering of x-rays by a single crystal of lysozyme. Unlike synthetic polymers, globular proteins may often be obtained as single crystals, because of their tight intramolecular packing (IQ ~ 1) and the resulting limited interactions between protein molecules in the solid state. It is generally believed that the solution and solid-state conformations and structures of globular proteins are quite similar, presumably as a consequence of predominant intramolecular interactions and packing.

The oxygen transport proteins myoglobin and hemoglobin are presented schematically in Figure 8.12, where their respective crystalline structures are shown together with the heme group that actually binds oxygen and carbon dioxide molecules. Hemoglobin binds oxy-

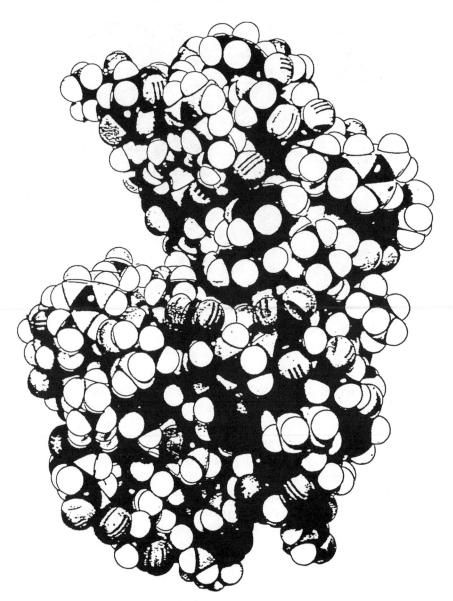

Figure 8.11. Molecular model of the three-dimensional structure of lysozyme (Dickerson and Geis, 1969).

(a)

(b)

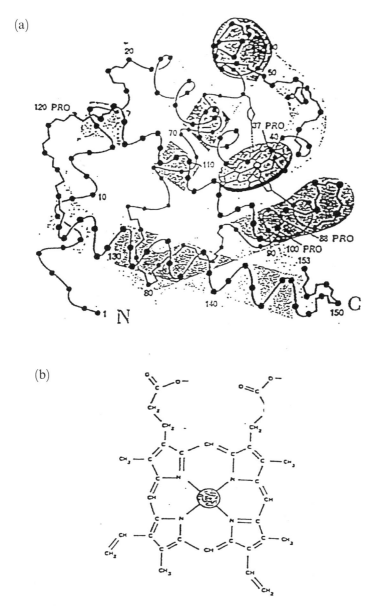

Figure 8.12. (a) Schematic representation of myoglobin three-dimensional structure, where each constituent amino acid residue has been reduced to a filled circle (Haynes and Hanawalt, 1968). (b) Skeletal structure of the heme group. (c) Same as (a) except replace myoglobin with hemoglobin (Dickerson and Geis, 1969).

(c)

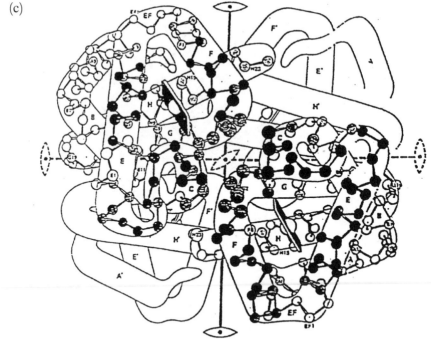

Figure 8.12. (continued)

gen in the lungs and delivers it to myoglobin in the muscles, where it is stored and then used in oxidative metabolic processes. The carbon dioxide byproduct of these metabolic processes is then transported by the myoglobin in the muscles to hemoglobin in the bloodstream for eventual exhalation from the lungs. Notice that myoglobin is locally α-helical, but with sufficient bends and breaks in and between helices to result in an overall globular shape. The 153 amino acid residues in the myoglobin chain serve primarily to fold this protein into a carrier for the heme group, which facilitates its ability to bind and release oxygen and carbon dioxide.

The hemoglobin protein is made of two pairs of polypeptide chains. The α-chains have 141 residues and the β-chains have 146 residues, and each α,β subunit carries a heme group. The conformation and overall shape and size of each subunit strongly resembles that of myoglobin, with a high α-helix content. The four subunit chains are packed so that extensive contacts occur between the side chains belonging to different subunits (α–β), but not between the same subunits (α–α, β–β). The packing of α and β subunits is symmetrical leading to an overall shape, which is roughly spherical. Not only can hemoglobin transport oxygen, but it can also control the amount of oxygen that binds to it.

Even more amazing than the sophisticated functions of proteins like myoglobin and hemoglobin is the fact that they are the direct result of their microstructures. The specific sequence of amino acid residues in a protein or the primary structures of a protein will alone determine its biological function. Though interesting, the amino acid composition of a protein does not determine its function. Rather it is the specific connectivity or sequence of amino acid residues composing the protein chain, which is the critical factor. Just as the physical properties of synthetic copolymers depend on their comonomer sequence distributions (random versus regularly alternating versus blocky), the primary structures of proteins establish their functions. For example, people who suffer with sickle-cell anemia have hemoglobin protein with a single mistake in its primary sequence (Watson et al., 1987). Normal hemoglobin has glutamic acid in the sixth position of its β-chains, but in those who suffer with sickle-cell anemia the glutamic acid has most commonly been replaced with valine (see Table 8.1). How can what appears on the surface to be a relatively minor structural change in hemoglobin's structure have such a remarkable effect on its function?

A likely structure–function scenario for hemoglobin follows: (1) The Val for Glu substition in the *primary structure* either partially disrupts or modifies the local conformation (i.e., the protein's *secondary structure*), in the vicinity of the sixth residue on the β-subunit chains. An α-helix is disrupted or the arrangment of neighboring α-helical regions is altered. (2) The overall folding or *tertiary structure* of the β-chains is modified by the change in local conformation. (3) The interactions, arrangements, and packing between the subunits (i.e., the *quaternary structure*) of the hemoglobin is affected by the altered folding and overall globular tertiary structure of the β-subunit chains. (4) The altered hemoglobin does not function effectively, because the dispositions and environments of its four heme groups have been modified, causing them to bind oxygen less effectively.

The complete hierarchy of protein structure from the *primary* (amino acid sequence) to the *secondary* (local conformation) to the *tertiary* (overall folding and shape) to the *quaternary* (arrangement and packing of subunits) is essentially controlled by the primary structure of the protein chain. Even if we are armed with knowledge of a protein's primary structure, we cannot currently determine its biological function, because we cannot as yet predict the resulting tertiary and/or quaternary structures (Grassberger et al., 1998), which in and of themselves might still not permit an understanding of the protein's biological function. Biomolecular scientists continue to isolate proteins and determine their structures in the hope of eventually being able to establish their structure–function relationships. Similarly, polymer scientists study the microstructures, conformations, overall sizes and shapes, and the solid-state packing and morphologies of synthetic polymers for the same reason—that is, to understand the bases of their unique properties.

Nature has 20 different amino acids from which to build proteins, so she does not lack for variety in constructing a wide range of proteins to meet a vast array of necessary biological functions. Rather, the problem of synthesizing a specific set of proteins unique to each living organism from the nearly limitless number of potential proteins, and doing so in a reproducible and efficient manner, is the issue that must be faced. In other words, how can an explicit set of specific proteins, each with a distinct amino acid residue sequence, be synthesized from a pool of 20 amino acids, which can potentially react with each other and themselves? For living organisms as simple as the single-cell *E. coli* bacterium, several thousand different proteins are nevertheless required to sustain life.

One possible approach would involve the random, step-growth synthesis of all possible proteins from the amino acid pool, followed by the application of a molecular sorting mechanism to select only the infinitesimally small subset of specific proteins required by each living organism. Even if we assumed that the 2500 proteins required by an organism, such as *E. coli*, contained only 100 amino acids each, of the $20^{100} \sim 10^{130}$ possible proteins only $(2500/10^{130}) = (2.5 \times 10^{-125})\%$ are necessary, and the rest must be discarded or depolymerized to refresh the amino acid pool. This highly inefficient scheme clearly is not possible for several reasons, among which is the sheer size requirement of such living organisms. Compared to an organism able to specifically synthesize only its required set of proteins, an organism employing random synthesis, followed by protein selection and recycling, would need to be roughly 10^{127} times larger. Can you imagine *E. coli* that are normally 20,000 Å long and 8000 Å thick growing to

3.2×10^{36} m in length and 1.3×10^{36} m in width to accommodate the random synthesis of its proteins? (Remember that the distances between Earth and the stars is measured in light years, where 1 light year is $\sim 10^{16}$ m. Even moving at the speed of light, a protein in this galactic *E. coli* would require $\sim 10^{20}$ years to travel from one end to the other end of the *E. coli* cell).

Instead, it is clear that a highly specific, directed, and efficient means must be available to living organisms by which they can reliably and reproducibly synthesize their required proteins. For over a century it has been known that hereditary information is transmitted from one generation of an organism to another by its cellular chromosomes. Each characteristic or trait of the organism is determined by particular portions or regions of the chromosomes called genes, which also control protein synthesis. Hence each gene is able to replicate itself so its genetic information is passed along to successive generations and it directs the synthesis of proteins necessary for the organism to function. This dual function of genes is central to life (Watson et al., 1987).

POLY(NUCLEIC ACIDS)

The fundamental constituents of genes and the associated mechanism for synthesizing proteins are provided by the poly(nucleic acids) DNA and RNA, which are presented in Figure 8.13. Both deoxyribonucleic acid (DNA) and ribonucleic acid (RNA) are polymers whose backbones are formed by regular repeating sugar units connected by phosphate bridges. Only D-ribose sugars are found in RNA and deoxy-D-ribose sugars (without the hydroxyl group on C2) in DNA. Notice that the phosphate bridges between sugar units are located at the $3'$ and $5'$ carbons, which in the pure sugars are bonded to hydroxyl groups that react with phosphorous acid to form the $-P-O-C_3$ and $-C_5-O-P-$ backbone links. Attached to each backbone sugar is an organic base, either a purine [adenine (A) or guanine (G)] or a pyrimidine [cytosine (C), thymine (T), or uracil (U)]. Note that T and U bases differ only by the presence or absence of a $-CH_3$ group, and only A, T, G, C bases are found in DNA, while only A, U, G, C bases are present in RNA.

Can we guess what key structural features of these poly(nucleic acids) give them the critical abilities to self-replicate and to contain or

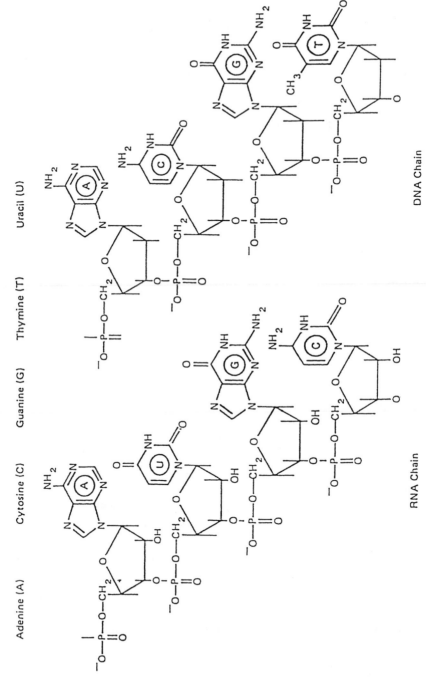

Figure 8.13. Skeletal structure of representative portions of RNA and DNA chains (after Mandelkern, 1972).

code the information necessary to direct the synthesis of a precise set of specific proteins? Because DNAs and RNAs both have homogeneous backbones—that is, single sugar types bridged by phosphate groups—we look to the organic bases A, C, G, T, U that are attached as side groups for our answer. DNA and RNA each have four distinct bases attached to their backbone sugar rings, so the composition and sequential distribution of attached bases are the only structural variables present. If the bases were organized in definite sequences along a DNA chain, then it is conceivable that this arrangement could be responsible for genetic coding. What would be the minimum base sequence length necessary to code for the synthesis of proteins?

Since proteins are synthesized from 20 different amino acids, there cannot be a $1:1$, base:amino acid code, because $4 \neq 20$. If a two-base sequence were the genetic code, then $4 \times 4 = 16$ and not 20 amino acids could be coded for. Clearly then, a three-base sequence in DNA would be more than sufficient $(4 \times 4 \times 4 = 64)$ to code for all 20 amino acids. This would require genetic material (DNA) sufficient to code for the totality of different proteins required by an organism. For example, if the one-cell *E. coli* bacterium required 2500 different proteins to function, and each had 100 amino acid residues, then *E. coli* DNA would have to consist of $3 \times 100 \times 2500 = 750,000$ structural units. In fact the molecular weights of DNAs have been found to range from several million to several tens of billions, depending on the complexity of the organism (Watson et al., 1987). These molecular weights correspond from several thousands to hundreds of millions of nucleotide repeating units. As a consequence, based on a three-letter code, there appears to be sufficient information in an organism's DNA to code and thereby direct the synthesis of all its proteins. Before presenting the details of the genetic code, let us consider how DNA can also be reliably replicated or reproduced, so that the transmission of identical genetic information from parent organism to offspring can be accomplished.

Comprehensive chemical analyses of DNA molecules from a variety of sources by Chargaff [as described in Crick (1962)] established the following base composition rules: (1) Number of A bases always equals number of T bases, (2) number of G bases always equals number of C bases, (3) number of A + G must therefore equal the number of T + C, and (4) number of $(A + T) \neq (G + C)$. These observations indicate that two distinct purine–pyrimidine base pairs, A-T and G-C, exist, because the amounts of the two bases making up a particular pair are the same, while the relative amounts of the two base pairs is variable. These DNA base pair relations were found universally, and so they pointed to the means by which DNA can replicate.

The ordered structure of DNA molecules, as observed and interpreted by Franklin and Gosling (1953), Wilkins et al. (1953), and Watson and Crick (1953), from x-ray scattering of oriented DNA fibers, provided the key information necessary to determine the means by which DNA can be replicated. Figure 8.14 presents a space-filling molecular model (a) and a schematic representation (b) of the now famous double-helical DNA structure first suggested by Watson and Crick (1953). Note that two complementary, right-handed helical chains that run in opposite directions are coiled around each other to form the DNA double helix, which has the appearance of a twisted ladder or spiral staircase. (An alternative view is provided in Figure 8.15, where the complementary nature of A-T and G-C base-pairing is shown.)

The sides or bannisters are composed of the sugar-phosphate backbones, and the rungs or steps consist of two bases, one from each chain, which lie in a plane perpendicular to the double-helical axis. The distance between chains is such that one of these bases must be a purine (A or G) and the other a pyrimidine (T or C). Two purines would not fit, while two pyrimidines would leave a gap between the chains. The Watson–Crick double-helical DNA structure fullfills the requirements that the numbers of A and T bases must be the same, the numbers of G and C bases must also be the same, but the numbers of $(A + T)$ and $(G + C)$ bases do not have to coincide. In other words, each base pair forming the rungs or stairs of the DNA double helix must be either A-T or G-C.

This base pairing scheme simultaneously satisfies both the chemical and steric requirements of DNA, while permitting abundant freedom in the distribution and overall content of bases, because $(A + T) \neq (G + C)$. A more detailed view of the complementary base pairs A-T and G-C is presented in Figure 8.16. Notice that the chain-to-chain separations are closely similar for both A-T and G-C base pairs, and the former is stabilized by two hydrogen bonds, while the latter benefits from three hydrogen bonds. The structural and chemical requirements of base-pairing assures that complementary A-T and G-C base pairs are formed between the DNA chains in the double helix and are key to the genetic function of DNA.

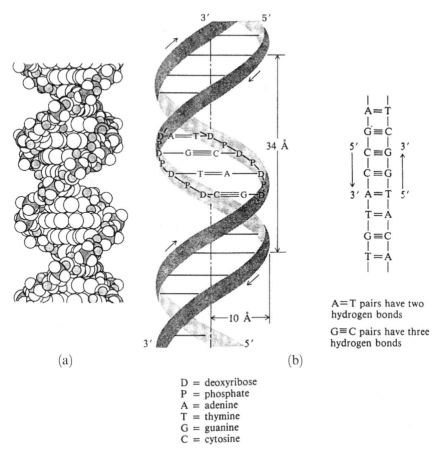

D = deoxyribose
P = phosphate
A = adenine
T = thymine
G = guanine
C = cytosine

Figure 8.14. Space-filling molecular model (a) and schematic representation (b) of the Watson–Crick (1953) double-helical structure of DNA (DuPraw, 1968).

Strict base-pairing between A-T and G-C bases belonging to the constituent double-helical chains of DNA means that the entire base sequence in one chain is complementary to the sequence of bases in the other chain. This is illustrated in Figure 8.15, where it can be seen that each chain in a sense may serve as a template for the other. During replication an exact copy of a mother cell's DNA must be transmitted to the new daughter cell. Since each of the complementary chains in the DNA double helix possess the information to specify the base sequence of the other, their separation would provide two templates for synthesizing exact complementary copies of each chain.

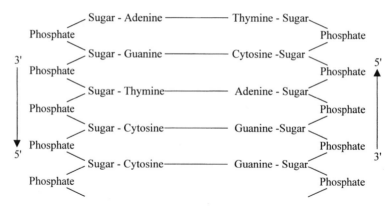

Figure 8.15. Schematic representation of the complementary nature of the two chains in the DNA double helix (after Mandelkern, 1972).

Such a process is diagrammed in Figure 8.17, where two identical daughter copies of the original parent DNA are produced. Although our description of DNA replication is admittedly simplified, the complementary nature of the two poly(nucleic acid) chains in the DNA double helix makes clear that each can serve as the synthetic template for the other, satisfying the critical need for DNA replication.

DNA-DIRECTED PROTEIN SYNTHESIS

Having briefly discussed its replication, let us return to the information that is encoded in DNA through its sequence distribution of bases. The message that an organism's DNA must deliver consists of the primary structure of each protein required by that particular organism. The four possible bases of DNA must code for the 20 possible amino acids in proteins, so, as we have mentioned, a three-letter base sequence XYZ, where X, Y, and Z are A, T, G, or C, is the minimum word length in the genetic code. Though the genetic code is contained and transmitted by DNA, RNA molecules under the control of DNA are the poly(nucleic acids) directly involved in protein synthesis. Actually three distinct types of RNA called messenger RNA (m-RNA), transfer RNA (t-RNA), and ribosomal RNA (r-RNA) are synthesized under the direction of DNA and carry the DNA-encoded information (m-RNA) and the amino acids (t-RNA) to the site of protein synthesis (r-RNA) in each cell. Messenger RNA receives the

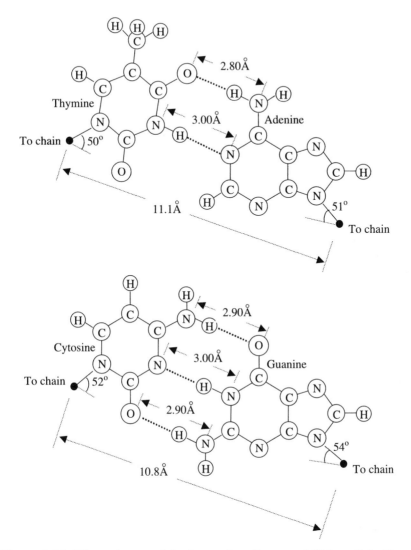

Figure 8.16. Dimensions and hydrogen-bonding capabilities of A–T and G–C base pairs in DNA (Anfinsen, 1960).

code from DNA and carries it to the ribosomes in the cell cytoplasm. There transfer RNA molecules each bound to a particular amino acid are marshaled in the correct order by m-RNA, a distinct t-RNA for each of the 20 amino acids, and they are polymerized to form the appropriate proteins (Watson et al., 1987; Frank, 1999).

The process of protein synthesis begins with the unraveling of

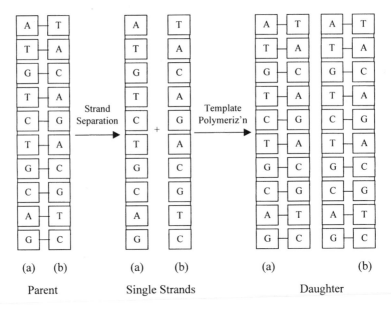

Schematic of Replication

Figure 8.17. Schematic diagram of DNA replication (Mandelkern, 1972).

DNA and the synthesis of a complementary m-RNA on the unwound DNA strand (see below). As an example, if the DNA sequence is C, C, T, G, G, T, A, A, C, then the complementary m-RNA sequence G, G, A, C, C, A, U, U, G is synthesized on the DNA template. The message is contained in successive three base groupings read along the complementary m-RNA, which does not assume a double-helical structure like DNA. A series of ingenious experiments using RNA molecules synthesized to have particular base triplets by Nirenberg (1961), Ochoa (1962), and Khorana (Jones et al., 1966), who then observed the primary structures of the resulting polypeptides, established the genetic code connecting the three base m-RNA sequences with each amino acid as presented in Table 8.2. Note there are also codons that signal the beginning(start) and termination(stop) of the synthesis of protein chains. The example 9-base sequence presented above for m-RNA would be expected, according to Table 8.2, to code for the tripeptide fragment -Gly-Pro-Leu-. There is a built-in redundancy in the genetic code, because many amino acids have more than one codon. This is useful protection against harmful mutations, which

TABLE 8.2. The Genetic Code; Translation of the Codons (m-RNA) into Amino Acids

First Base (5' end)	Second Base	Third Base (3' end)			
		U	C	A	G
U	U	Phe	Phe	Leu	Leu
	C	Ser	Ser	Ser	Ser
	A	Tyr	Tyr	Stop	Stop
	G	Cys	Cys	Stop	Trp
C	U	Leu	Leu	Leu	Leu
	C	Pro	Pro	Pro	Pro
	A	His	His	Gln	Gln
	G	Arg	Arg	Arg	Arg
A	U	Ile	Ile	Ile	Met (start)
	C	Thr	Thr	Thr	Thr
	A	Asn	Asn	Lys	Lys
	G	Ser	Ser	Arg	Arg
G	U	Val	Val	Val	Val
	C	Ala	Ala	Ala	Ala
	A	Asp	Asp	Glu	Glu
	G	Gly	Gly	Gly	Gly

can alter or transform the sequence of nucleic acids in DNA. Also note that neither hydroxyproline or cystine have code words, so they must be synthesized from their close relatives proline and cysteine after the protein chain is formed.

The amino acids are bound to t-RNA molecules, which are small (70–80 nucleotdes) and possess unusual structures (see Figure 8.18). Importantly, each has a different amino acid attached and an anti-codon, or three base sequence, corresponding to that amino acid. Because m-RNA has a base sequence complimentary to DNA, and t-RNA must bind to m-RNA, the anti-codon on t-RNA and the DNA base sequence from which the m-RNA was synthesized must be identical. Figure 8.19 presents a highly idealized visualization of protein synthesis, where the functions of m- and t-RNA are empha-sized. A schematic diagram describing the roles generally played by poly(nucleic acids) in biological functions is presented in Figure 8.20, where the relationship between m- and r-RNA molecules is

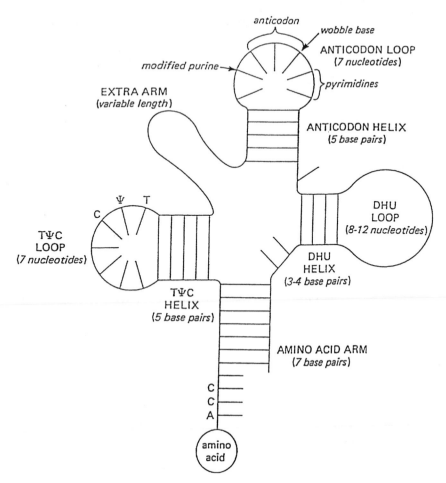

Figure 8.18. Cloverleaf diagram of the secondary structure for all t-RNA molecules (Arnott, 1971). Critical features distinguishing among the t-RNA molecules are the specific anti-codon and bound amino acid.

also indicated. Protein biosynthesis also requires scores of enzymes and other molecules, which we have not mentioned; however, the poly(nucleotides), DNA and RNAs, are central to the specificity and efficiency of the process. It is estimated (Hart et al., 1995) that a protein containing as many as 150 amino acid residues can be assembled biosynthetically in less than a minute. For synthetic chemists, this is indeed a humbling observation.

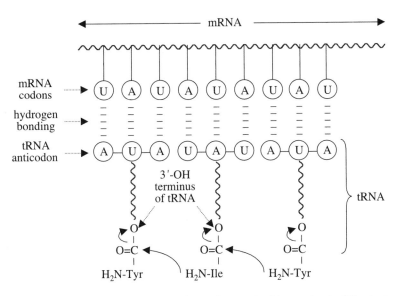

Figure 8.19. Schematic representation of protein biosynthesis (Hart et al., 1995).

We conclude this chapter on biopolymers with a brief introductory discussion of several additional features of proteins and poly(nucleic acids).

PROTEIN FOLDING

The DNA-directed and RNA-mediated synthesis of proteins on the ribosomes of cells produces an exact set of proteins, each with a precise primary structure. Some of these proteins are fibrous with a structural function and have regular repeating secondary structures leading to all *trans*, planar zigzag or helical shapes and extended sizes. Other proteins are highly folded into compact, globular, yet precise shapes or tertiary structures and perform a variety of chemical, transport, and regulatory functions for an organism. The question arises: What microstructural elements in the primary structures of proteins lead to specific local conformations or secondary structures, which, when sequentially coupled along the protein chain, result in a particular overall size and shape or tertiary structure? (Grassberger

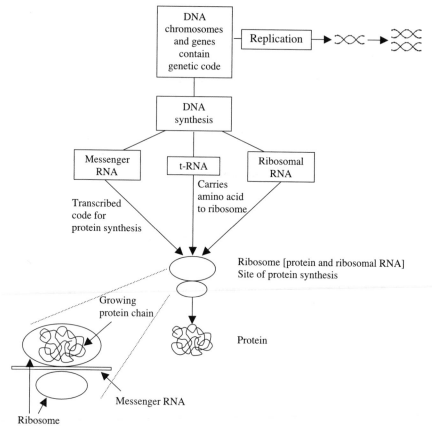

Figure 8.20. Summary diagram of role of nucleic acids in biological function (Mandelkern, 1972).

et al., 1998). To put the following discussion into context it must be said that presently we are not able to predict the detailed tertiary structure of a protein solely from knowledge of its primary structure. We can, however, make suggestions conconcerning ways in which a protein's microstructure can influence how it is folded and stabilized into its biologically active tertiary structure, which determines its living function.

To begin with, the inherent conformational characteristics of each amino acid residue in a protein depend on the nature of its particular side group (see Table 8.1). For example, compare the conformational energy maps for the glycine and L-alanine residues presented in Figure 8.6. Proline provides another example, because its $-N-C^\alpha-\phi$ rotation angle is restricted by the pyrrolidine ring to $\sim 120°$. In addition, the

peptide amide bond ($-C-\overset{\overset{\displaystyle O}{\|}}{C}-\underset{\underset{\displaystyle CH_2^-}{|}}{N}-C-$) is no longer required to be

trans with respect to the backbone C^α carbons in the proline and preceding residues (see Figure 8.21), because two sp^3 carbons are bound to the proline amide nitrogen. Figure 8.21 also presents the conformational energy map for an isolated L-prolyl residue (b), such as the one seen in (a). A comparison with Figure 8.6 makes clear that L-proline residues are conformationally restricted compared to glycine and L-alanine residues, for example.

In addition to the limited conformations available to an L-prolyl residue, the conformations of residues immediately preceding L-proline in a protein are also restricted by its presence, as evidenced in Figure 8.22. Here the conformational energy maps for glycine (a) and L-alanine (b) residues immediately followed by L-proline are presented for both *trans* and *cis* connecting peptide bonds—that is, the $C_{i-1}-N_i$ bond in Figure 8.21(a). Comparison with the Gly and L-Ala energy maps in Figure 8.6 makes obvious that the L-proline amino acid residue seriously constrains the conformations available to the previous amino acid in a protein primary structure, especially when connected by a *cis* peptide bond. Judicious placement of L-Pro residues in the primary structure of a protein can clearly be expected to have important consequences. For example, L-Pro can only occur at the beginning of an α-helical region in a protein, because the region of the right-hand α-helical conformation around $\phi, \psi = 120°$, $120°$ is not accessible to any immediately preceding amino acid residue except Gly when connected by a trans peptide bond (see Figure 8.22). For this reason, L-Pro is often called an α-helix breaking amino acid residue.

Poly-L-proline adopts a left-handed helical conformation with $\psi = 300-330°$ (see Figures 8.10 and 8.21). In collagen (the protein found in cartilage, ligaments, tendon, and the matrix upon which bone mineralizes), the amino acid triplets Gly-X-Pro, Gly-X-Hypro, and Gly-Pro-Hypro dominate its primary structure, and so it may not be too surprising that collagen also adopts the left-handed poly-L-proline helical conformation seen in Figure 8.10(a). As mentioned earlier,

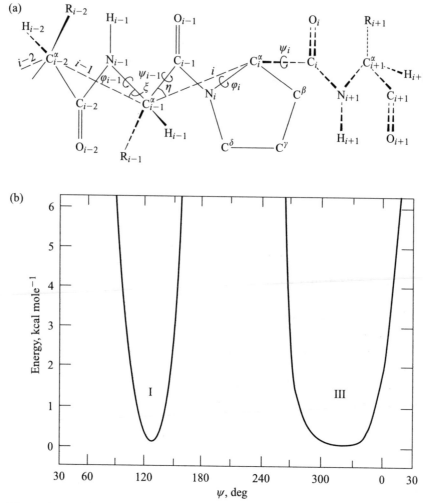

Figure 8.21. (a) Schematic diagram of a protein chain containing an L-prolyl residue, where $\phi \sim 120°$ is enforced by the pyrrolidine ring and this moves all atoms to its right out of the plane of the atoms to its left (Flory, 1969). (b) Conformational energy E (ψ) calculated for an isolated L-prolyl residue, with $\phi = 122°$ (Schimmel and Flory, 1968).

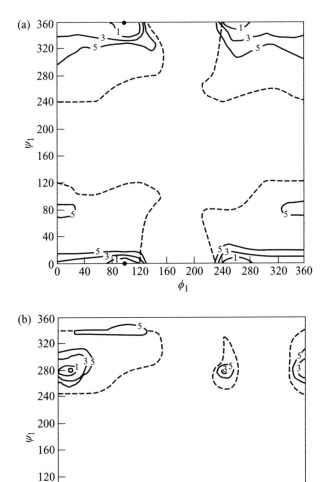

Figure 8.22. Conformational energy map for a glycine (a) and L-alanine (b) residue preceding L-Pro ($\phi = 120°$) with a *cis* peptide bond (Tonelli, 1974a). Energy contours in kcal/mol relative to the lowest energy conformer denoted by an X. The dashed lines are the 5.0-kcal/mole contours obtained for Gly (a) and L-Ala-L-Pro (b) by Schimmel and Flory (1968) with a *trans* connecting peptide bond.

three collagen protein chains coil together to form tropocollagen [Figure 8.10(b)]. In the tropocollagen superhelix, two of every three amide (NH) and carboxyl ($-\overset{\overset{\text{O}}{\|}}{\text{C}}-$) groups in each collagen chain hydrogen-bond to one of the other two chains as indicated by the dashed lines in Figure 8.10(b). Because of the distances between chains and the periodicities of their helical conformation, amino acid residues with side chains cannot be accommodated in every third repeating unit, and thus Gly must occupy this position. The necessity of Gly-X-Pro or Hypro amino acid triplets in the primary structure of collagen is quite evident, but only after the fact.

We could have anticipated the left-handed poly-L-proline helical conformation for collagen, since nearly two-thirds of its amino acid residues are Pro and Hypro, but without knowledge of the tropocollagen structure it is not readily apparent why Gly appears in every third amino acid residue position. This illustrates an important feature of fibrous or structural proteins with regular repeating and extended conformations. Because they are generally organized into structural units containing more than a single protein chain (as in tropocollagen, the protofibrils of α-keratin, and the β-sheets of the nearly fully extended silk proteins), the primary structures of fibrous proteins must not only lead to the correct tertiary structure or regular repeating conformation, but must also meet the requirements imposed by packing the extended protein chains into larger organized units or quaternary structures. This latter requirement, therefore, continues to make complete understanding of the function of fibrous proteins solely from knowledge of their primary structures a formidable challenge.

Another property specific to an amino acid that has important consequences for the folding of proteins is associated with the cysteine residue (see Table 8.1). The $-CH_2-SH$ side chains of two proximal CySH residues that are properly aligned may form a $-CH_2-S-S-CH_2-$ disulfide bond converting the two CySH residues to cystine (Cys). Cys molecules, which chemically bond or form internal cross-links between CySH residues, clearly function to stabilize the tertiary structures of proteins. For example, the globular protein lysozyme, which catalyzes the degradation of cell walls, contains 8 CySH residues that form 4 Cys disulfide links as indicated in Figure 8.23. These internal Cys disulfide links serve to stabilize the tertiary structure of

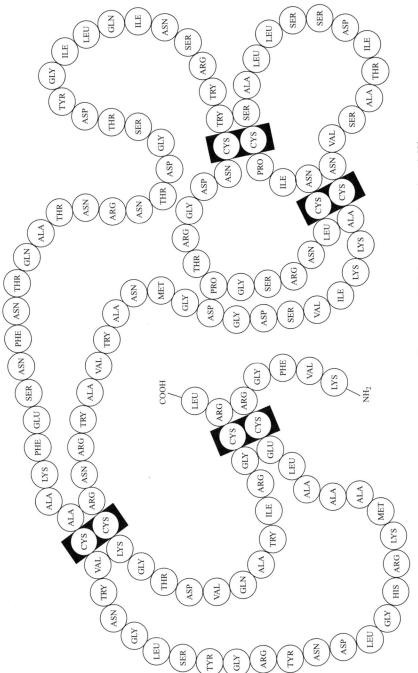

Figure 8.23. Primary structure of lysozyme (Dickerson and Geis, 1969).

lysozyme, shown in Figure 8.11, and ensure that its catalytic activity is maintained. In addition, sequential formation of the Cys disulfide links probably assists the folding of lysozyme into its active tertiary structure.

Cys disulfide links are not limited to the bonding of CySH residues within a protein, but may also stabilize structural elements or quaternary structures important to the functioning of fibrous proteins. For example, the α-keratin protofibrils, consisting of a three-stranded rope of α-helical, α-keratin chains (see Figures 8.8 and 8.9), are stabilized by Cys disulfide links formed between the α-keratin chains in each protofibril. Mild reducing agents can disrupt the $-CH_2-S-S-CH_2-$ links while mild oxidizing agents can form them from CySH residues, thereby becoming the basis for permanent waving of hair. Application of a mild reducing agent to hair disrupts the Cys links and allows it to be rewaved. The rewaved hair is then treated with a mild oxidizing agent to reform Cys links between different CySH residues, thereby stabilizing the new hairdo. Of course as hair grows and new α-keratin replaces old, the "hairdo" reverts to the former, natural look, because of the natural distribution of Cys disuldide links in the freshly biosynthesized and organized α-keratin protofibrils. Permanent waving of hair is only as permanent as the rate of hair growth.

Inspection of Table 8.1 reveals that several amino acids possess side chains with polar groups capable of forming hydrogen bonds with the side chains of the same or other amino acids. Consequently, just as Cys links formed between two CySH residues that are appropriately juxtaposed can assist the folding and stabilize the tertiary structure of the folded protein, so too may these dipolar and/or hydrogen-bonding interactions between amino acid side chains that are proximal in space, but distant along the primary sequence. The following statements serve as challenges to those who are still attempting to learn how proteins fold and why only a single-folded protein tertiary structure results from a specific primary structure: (1) Proteins are polypeptide chains generally composed of a 100 or more amino acid residues of which there are 20 different kinds, (2) most amino acid residues are inherently conformationally diverse, though L-Pro and those residues immediately preceding L-Pro are conformationally restricted, (3) the side chains in several amino acid residues can interact with each other to varying degrees spanning the range from van der

Waals to hydrogen bonding to formation of chemical –S–S– bonds, and (4) the folded or tertiary structures of proteins, whether fibrous or globular, are internally close-packed with IQs → 1. It is not surprising that a problem like protein folding, involving so many atoms and conformational degrees of freedom, has resisted solution for over 50 years.

DNA REPLICATION AND TRANSCRIPTION

Two related questions concerning the replication and transcription of DNA may be raised. First, which of the two DNA chains in the double helix codes for and directs protein synthesis, and second, is the m-RNA that functions as the template for protein synthesis a complementary copy of only one of the two strands of the DNA double helix? Because the two strands of double-helical DNA are complementary and not exact copies, while the genetic code is not complementary— that is, UUU, CUA code for Phe, Leu, but the complementary triplets AAA, GAU code instead for Lys, Asp—each potentially contains distinct information, leading to synthesis of different sets of proteins. Though it is known that proteins are synthesized under the control of only one of the two double-helical DNA strands, the mechanism for descriminating between strands is not yet understood (Watson et al., 1987; Vol'kenstein, 1970) What is clear is the importance of DNA strand separation in both DNA replication and transcription of DNA information to m-RNA, as indicated in Figure 8.24.

Even in viral organisms that possess their genetic information on single strands of DNA, strand separation is crucial to their replication, because as soon as they enter a host cell a complementary copy of their single DNA strand is synthesized, followed by normal replication as indicated in Figure 8.24. These viruses only make proteins corresponding to the original single strand of DNA, just as only one strand of double-helical DNA in higher organisms directs protein synthesis. How this choice is made, however, is not currently understood. As indicated in part (b) of Figure 8.24, m-RNA is only synthesized on one of the two double-helical DNA strands, but how the remaining strand is prevented from transcribing its genetic information to m-RNA remains a mystery.

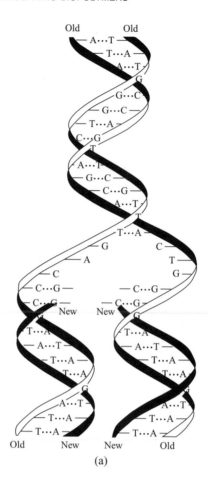

(a)

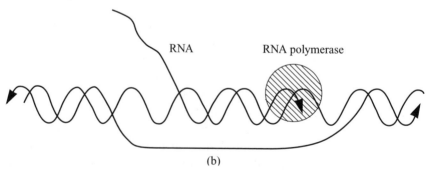

(b)

Figure 8.24. (a) Replication of DNA. (b) Transcription of m-RNA molecule upon a unique strand of DNA (Watson et al., 1987).

THE GENETIC CODE

The DNA-directed three-base sequence code transcribed and delivered by m-RNA to incorporate the 20 amino acids into the appropriate protein primary structures was established by studying the polypeptides synthesized from regularly repeating polyribonucleotides. For example, the tetranucleotide UAUC was polymerized and then subjected to peptide-synthesizing conditions. The regularly repeating polypeptide (-Tyr-Leu-Ser-Ile-)$_n$ was obtained. The coding polyribonucleotide ...UAUCUAUCUAUCUAUC... possesses the following repeating three base codons: UAU, CUA, UCU, and AUC. Hence, we may associate UAU and Tyr, CUA and Leu, UCU and Ser, and AUC and Ile (see Table 8.2). Similarily, the polyribonucleotide poly (UA) generated the polypeptide poly(Tyr-Ile). The three-base sequence UAU, as expected, incorporated Tyr, while AUA produced Ile. Thus, we see from these examples that both AUA and AUC code for Ile, a redundancy that is observed for most of the amino acids (see Table 8.2) and serves as a safeguard against certain genetic mutations.

Genetic mutations caused by radiation, cancer-producing agents, or other means generally replace one DNA base with another or add or delete a base. If, for example, the UUU base triplet were mutated to UCU, Ser instead of Phe would be inserted into the protein. However, if UCU mutated to UCC, Ser would still be inserted into the protein chain. Clearly, having some redundancy built into the genetic code can serve to at least partially protect its integrity against genetic mutations. Unlike single base substitution mutations, when a nucleotide and its associated base is added or deleted from DNA the consequences are more serious. Substitution can only potentially cause the insertion of a single incorrect amino acid into the protein chain, while the addition or deletion of a nucleotide causes a whole scale shift in the reading of the genetic message (Watson et al., 1987; Vol'kenstein, 1970).

Suppose DNA containing the nucleotide sequence ...(ATA)· (CGA)(TTC)(GTG)..., which produces the m-RNA sequence ...(UAU)(GCU)(AAG)(CAC)... and codes for the ...(Tyr)(Ala)· (Lys)(His)... protein sequence, added a C nucleotide between the G and A units during mutation. The mutated DNA sequence would now read ...(ATA)(CGC)(ATT)(CGT)G..., with the transcribed m-RNA sequence ...(UAU)(GCG)(UAA)(GCA)C.... Because GCU and GCG are redundant for Ala, the first two amino acids incorporated into the protein chain would remain Tyr and Ala; however, the third codon in the mutated DNA is now UAA, which is the code for stopping the synthesis of the current protein (see Table 8.2 and note that there is also a a codon AUG that begins the synthesis of each distinct protein). Mutations that add or delete a nucleotide from DNA can generally be expected to alter the amino acid coded for at the site of mutation and also most of the subsequent amino acids incorporated, because of the translational shift produced in its genetic message.

When a virus infects a host it is able to reproduce by making its genetic information accessible to the host cell's replication and protein synthesis chemistry. The host cell then reproduces not only itself, but the infecting virus as well. By modifying this process, an organism may be genetically engineered to produce proteins that it normally would not. This is typically made possible by splicing the genetic information required for the synthesis of the desired protein into the DNA of the host organism. For example, if the desired protein consists of 100 amino acids, then DNA with 306 nucleotides, beginning and ending with start and stop codons tied together with the 100 codons that correspond to the primary structure of the protein, must be spliced into the host's DNA. Once gene-splicing enzymes were identified (Dawson et al., 1996), it became possible to genetically engineer simple organisms like bacteria and plants to achieve the synthesis of proteins desired by the "gene-splicer" in a surrogate organism that for practical reasons, such as abundance, size, and ease of growth, is preferred over the organisms that naturally produce the desired proteins.

An example of gene-splicing, genetic engineering comes from the recent announcement that the gene for producing silk in a spider has been spliced into the mammary cells of a goat (Chemistry ACS, p. 4, Winter 1999). Previously, bacteria have been engineered to produce spider silk protein, but its Cys-Cys, interprotein cross-linking was disordered resulting in a "lumpy mess." Because mammary cells are more similar to the silk-producing cells of spiders, the silk protein synthesized in the goat bond together in a more orderly fashion. However, it remains to mimic the spider's fiber spinning talents before goats can be milked for silk fibers, as well as for their milk.

Recently, efforts to determine the genetic information encoded in the DNAs of a wide range of organisms by rapid sequencing techniques has greatly expanded. The most ambitious effort among these is the Human Genome Project (Bodmer, 1995), whose goal is to completely sequence the entire collection of human DNA. To offer some perspective on this effort, it must be remembered that the single-cell bacterium *E. coli* has 4 DNAs, which synthesize some 2000 to 3000 different proteins (Watson et al., 1987). It has been estimated that the human genome may contain 1000 times more information. The latest estimate is that within a year's time we will possess an Atlas of the human genome (Chemistry ACS, Winter 1999). This will enable the complete cataloging of human proteins in terms of their primary structures, which is of course a basic necessity for eventually determining a detailed description of human biology. Without knowledge of its basic constituents, it is difficult to begin to understand how an organism functions.

The genetic information carried by DNA resides in the sequence of its constituent nucleotides. To learn which proteins are encoded in the DNA of an organism, we must be able to "read" or sequence the primary structure of its DNA. Sequencing of DNAs to reveal their genetic information—that is, to learn which sets of specific proteins they will synthesize—has revolutionized molecular biology (Sanger and Dowding, 1996; Chirikjian, 1981). Instead of the laborious procedure of isolating, purifying, and sequencing the complete set of an organism's constituent proteins, sequencing its DNA and applying the genetic code provides the same information directly.

We close with a practical application of genomic science. Roughly 5% of human DNA codes for protein enzymes, hormones, transport agents, and other peptides that are identical and do not vary from person to person. However, the remaining DNA, which generally code for structural proteins, varies tremendously among individuals, and, except for identical twins, is characteristic of an individual. An individual can be identified by the DNA carried in his or her genes, and DNA profiling or "fingerprinting" has increasingly found use in forensic science as a powerful means of personal identification (Hart et al., 1995).

A small quantity (as little as a few micrograms) of DNA obtained from blood, hair, skin, nails, saliva, semen, or other tissue samples usually suffice for analysis. The DNA is purified and cut with restriction enzymes, which cleave DNA at known four-base sequences, to produce smaller fragments of $\sim100-150$ nucleotides in length. These fragments are separated and labeled at the 5′ end with radioactive phosphate containing ^{32}P. The labeled fragments are then further degraded by subjecting them to four carefully controlled reaction conditions and specific reagents, which splits their chains at a particular base, A, T, G, or C. This results in a group of DNA pieces containing one, two, three, and more nucleotides, each ^{32}P-labeled. The products of these four parallel experiments are then separated by gel-electrophoresis (Gaal et al., 1980).

Poly(acrylamide) is coated on a glass plate and then cross-linked and swollen with a salt solution. The small DNA fragments are segregated at the bottom of the gel in four separate zones one corresponding to each of the A, T, C, and G cleaved products. A voltage is applied across the gel from bottom to top, causing each of the species in each zone to migrate distances that depend on the number of nucleotides it contains. This oligonucleotide separation is produced when the charged nucleotides (Note the negative charge on each phosphate unit in Figure 8.13) migrate under the influence of the applied electric field, and the distances to which they migrate are proportional to the number of charges or nucleotides they contain. The DNA sequence can be read directly from the positions of the radioactive spots on the gel-electrophoresis plate.

The pattern of radioactively labeled DNA fragment spots is unique for each human being, except for identical twins. In addition to its forensic applications, these and additional rapid gene-sequencing techniques are being employed in the Human Genome Project to eventually unravel the complete humane genome containing several billion base pairs. Various genetically related diseases have also been investigated by these means. Any irregularities detected in the DNA profile of that part of the human genome ($\sim 5\%$) that is normally invariant and controls the synthesis of globular proteins signal a genetic defect that may manifest a disease. Sequencing the defective gene yields the primary structures of the proteins whose syntheses it directs, and these proteins (or DNA sequences) may be compared to the normal complement to pinpoint the source of a genetically related disease.

Following our discussion of the proteins and poly (nucleic acids), which manifest a host of complex structures and sophisticated functions that are crucial for life, it may occur to the reader to ask whether or not nonpolymeric material could suffice to sustain life. In other words, are macromolecules necessary constituents of living organizms? As pointed out recently by DiMarzio (1999), the answer is very likely yes, because self-assembly is expressed most fully in polymeric systems whose constituents, the individual polymer chains, can greatly alter their structures as a consequence of unique intramolecular phase transitions. For example, proteins can randomly coil when denatured, but are also able to fold into highly extended or closely packed, globular conformations, with regular repeating and non-repeating, yet specific, conformations, respectively. After synthesis on the cell's ribosome, an α-keratin chain may be transported as a randomly coiling protein chain in the cells found in the cuticle or hair follicles, where environmental conditions are conducive to producing an intramolecular random-coil to α-helix conformational transition. Three α-helical, α-keratin chains may then entwine to form the three-stranded α-keratin protofibril, which is further stabilized by inter-strand Cys disulfide bonds. This latter step and the final organization of α-keratin protofibrils (see Figures 8.8 and 8.9) into hair or cuticle are primarily intermolecular events, but each could not occur without the intramolecular random-coil to α-helix transition, which has no counterpart in atomic or small-molecule systems.

It is, once again, the long-chain nature of polymers which confers upon them the ability to greatly alter their sizes and shapes in response to their environments. In this intramolecular manner they are able to begin the process of self-assembly into complex biological structures composed of many protein chains, such as hair, or single globular proteins able to catalyze chemical reactions or transport vital small molecules in living organisms.

DISCUSSION QUESTIONS

1. What inherent property of both the cellulose and amylose chains might explain their different solubilities and thermal properties? Also examine their structures (Figure 8.2) and discuss possible differences in their abilities to pack intermolecularly.

2. Suggest a possible reason why replacement of the $-\overset{\overset{\displaystyle O}{\|}}{\underset{\underset{\displaystyle H}{|}}{N}}-C-CH_3$ group in chitin with the $-NH_2$ group in chitosan leads to solubility.

3. Compare and contrast the expected inherent conformational flexibilities of polypeptides and proteins to aliphatic polyamides such as nylons-6 and -6,6. Consider both their conformationally averaged dimensions, as characterized by C_n, and the conformational accessibility of each of their backbone bonds. (Note that $C_n = 10$ and 6 for polypeptides like poly-L-alanine and nylon-6 and -6,6, respectively, where nl^2 is the number of L-alanine residues times the square of the virtual bond connecting C^α carbons and the number times the average square of the backbone bond lengths in the nylons.)

4. The following small protein Ala_1-Gly_2-Ser_3-Ala_4-Pro_5-Phe_6-Gly_7-His_8-Val_9-Gly_{10} adopts an overall U-shape, where the Ala_1-Gly_2-Ser_3-Ala_4 and Gly_7-His_8-Val_9-Gly_{10} residues are α-helical and form the legs of the U and the Pro_5 and Phe_6 residues are in a folded conformation and form the bend in the protein U-shape. Describe the primary, seconary, and tertiary structures of this protein. Also discuss anything that might appear suspicious about its local conformations.

5. Characterize the similarities and differences between the behaviors of randomly-coiling synthetic polymers and the globular proteins. Which have a larger entropy and why, and which appear to be dominated by long-range interactions that do not lead to chain expansion and why?

6. Describe how the complementary, double-helical structure of DNA serves to make it reliably reproducible during its replication. Can you suggest how single-stranded DNA might be replicated?

*7. Contrast the inherent conformational characteristics of the Gly, L-Ala, and L-Pro amino acids in a protein. Can it be said that one is more or less flexible than the other, and if so, why?

*8. Discuss the role of L-Pro at the fifth position in the protein of Question 4. Is it resonable that L-Pro occupies this position considering the tertiary structure of the protein?

*9. In addition to stabilizing the overall tertiary and quaternary structures of proteins, can you suggest how Cys-Cys–CH_2–S–S–CH_2– cross-links might be involved in protein folding?

*10. Is a point or addition/deletion mutation of DNA potentially more serious and why?

*11. Can you suggest how as little as a few micrograms of DNA can be easily sequenced?

*12. Why would the comparison of DNAs of healthy people with those from individuals suffering a disease associated with the malfunctioning of structural or fibrous proteins be difficult to use for establishing the source of the malfunctioning protein?

REFERENCES

Anfinsen, C. B. (1960), *The Molecular Basis of Evolution*, John Wiley and Sons, New York.

Arnott, S. (1971), *Prog. Biophys. Mol. Biophys.*, **22**, 181.

Bell, J. E., and Bell, E. (1988), *Proteins and Enzymes*, Prentice-Hall, Englewood Cliffs, NJ.

Bodmer, W. F. (1995), *The Book of Man: The Human Genome Project and the Quest to Discover our Genetic Heritage*, Scribners, New York.

Brant, D. A., and Christ, M. D. (1990), *Computer Modeling of Carbohydrates*, American Chemical Society, ACS Symposium Series #430, Washington, D.C., Chapter 4.

Brant, D. A., and Flory, P. J. (1965), *J. Am. Chem. Soc.*, **87**, 663, 2788, 2791.

Brant, D. A., Miller, W. G., and Flory, P. J. (1967), *J. Mol. Biol.*, **23**, 47.

Chirikjian, J. G. (1981), *Restriction Endonucleases*, Elsevier/North Holland, New York.

Cowan, P. M., and McGavin, S. (1955), *Nature*, **176**, 501.

Crick, F. H. C. (1962), *The Genetic Code*, Scientific American, October.

Daniel, J. R. (1990), *Concise Encyclopedia of Polymer Science and Engineering*, Kroschwitz, J. L., Ed., John Wiley and Sons, New York, p. 124.

Dawson, M. T., Powell, R., and Gannon, F. (1996), *Gene Technology*, Biosci Science Publishers, Oxford, United Kingdom.

Dickerson, R. E., and Geis, I. (1969), *The Structure and Action of Proteins*, Harper and Row, New York.

DiMarzio, E. A. (1999), *Prog. Polym. Sci.*, **24**, 329.

DuPraw, E. J. (1968), *Cell and Molecular Biology*, Academic Press, New York.

Filshie, B. K., and Rogers, G. E. (1961), *J. Mol. Biol.*, **3**, 501.

Flory, P. J. (1969), *Statistical Mechanics of Chain Molecules*, Wiley-Interscience, New York.

Frank, J. (1999), *Sci. Am.*, **86**, 428.

Franklin, R. E., and Gosling, R. G. (1953), *Nature*, **171**, 740, 4356; **172**, 156.

Gaal, O., Medgyesi, L., and Vereczkey, L. (1980), *Electrophoresis in the Separation of Biological Macromolecules*, John Wiley and Sons, New York.

Grassberger, P., Nadler, W., and Barkema, G. T. (1998), Eds., *Monte Carlo Approaches to Biopolymer and Protein Folding*, World Science Press, River Edge, NJ.

Hackin, P. (1980), *The Sizing of Spun Yarns*, Scientific ERA Publishers.

Hart, H., Hart, D. J., and Crane, L. E. (1995), *Organic Chemistry: A Short Course*, Houghton Mifflin, Boston.

Haynes, R. H., and Hanawalt, P. C. (1968), *The Molecular Basis of Life*, W. H. Freeman, San Francisco.

Hirano, S., Hayashi, M., Nishida, T., and Yamamoto, T. (1989), *Chitan and Chitosan*, Skjak-Braek, G., Anthonsen, T., and Sanford, P., Eds., Elsevier, London, p. 743.

Holum, J. R. (1994), Fundamentals of General, Organic, and Biological Chemistry, John Wiley and Sons, New York.

Jackson, C., and O'Brien, J. P. (1995), *Macromolecules*, **28**, 5975.

Jones, D. S., Nishimura, S., and Khorana, H. (1966), *J. Mol. Biol.*, **16**(2), 454.

MacGregor, E. A., and Greenwood, C. T. (1980), *Polymers in Nature*, John Wiley and Sons, New York.

Mandelkern, L. (1972) "An Introduction to Macromolecules", Springer-Verlag, New York.

Marsh, R. E., Corey, R. B., and Pauling, L. (1955), *Biochem. Biophys. Acta*, **16**, 13.

Matheja, J., and Degens, E. T. (1971), *Structural Molecular Biology of Phosphate*, Fischer, Stuttgart, Germany.

Mathur, A. B., Tonelli, A. E., Rathke, T. D., and Hudson, S. M. (1997), *Biopolymers*, **42**, 64.

Natta, G., and Corradini, P. (1960), *Nuova Cimento., Suppl.* **15**, 1, 9, 40.

Nirenberg, M. W., and Matthaei, J. H. (1961), *Proc. Natl. Acad. Sci. USA*, **47**, 1580, 1588.

Ochoa, S., et al. (1962), *Proc. Natl. Acad. Sci. USA*, **47**, 1936.

Okano, T., and Sarko, A. (1984), *J. Appl. Polym. Sci.*, **29**, 4175.

Pauling, L. (1960), *The Nature of the Chemical Bond*, 3rd ed., Cornell University Press, Ithaca, NY.

Pauling, L., and Corey, R. B. (1953), *Nature*, **171**, 59.

Rathke, T. D., and Hudson, S. M. (1994), *J. Macromol. Sci.*, **C34**(3), 375.

Rice, R. V. (1960), *Proc. Natl. Acad. Sci. (US)*, **46**, 1186.

Rorrer, G., Hsien, T., and Way, J. D. (1993), *Ind. Eng. Chem. Resources*, **33**(9), 2170.

Sanger, F., and Dowding, M. (1996), *Selected Papers of Frederick Sanger*, World Scientific, River Edge, NJ.

Sarko, A. (1976), *Appl, Polym. Symp.*, **28**, 729.

Sasisekharan, V. (1959), *Acta Crystallogr.*, **12**, 897.

Schimmel, P. R., and Flory, P. J. (1968), *J. Mol. Biol.*, **34**, 105.

Tonelli, A. E. (1974a), *J. Mol. Biol.*, **86**, 627.

Tonelli, A. E. (1974b), in *Analytical Calorimetry*, Vol. 3, Porter, R. S., and Johnson, J. F., Eds., Plenum, New York.

Tonelli, A. E. (1986), in *Cyclic Polymers*, Semlyen, J. A., Ed., Elsevier, London, Chapter 8.

Tokura, S., Hasegawa, O., Nishimura, S., Nishi, N., and Takatori, T. (1987), *Anal. Biochem.*, **161**, 117.

Turbak, A. (1990), in *Concise Encyclopedia of Polymer Science and Engineering*, Kroschwitz, J. L., Ed., John Wiley and Sons, New York, p. 960.

Vol'kenstein, M. V. (1970), *Molecules and Life*, translated from the Russian by Timasheff, S. N., Plenum, New York.

Watson, J. D., and Crick, F. H. C. (1953), *Nature*, **171**, 737, 964, 4356, 4361.

Watson, J. D., Hopkins, N. H., Roberts, J. W., Steitz, J. A., and Weiner, A. M. (1987), *Molecular Biology of the Gene*, 4th ed., Benjamin-Cummings, Menlo Park, California.

Wilkins, M. H. F., Seeds, W. E., Stokes, A. R., and Wilson, H. R. (1953), *Nature*, **172**, 759, 4382.

INDEX

RENNER LEARNING RESOURCE CENTER
ELGIN COMMUNITY COLLEGE
ELGIN, ILLINOIS 60123